# Civil Engineering Construction
# Design and Management

## Other titles of interest to Civil Engineers

*Soil Mechanics – Principles and Practice, Second edition*
G. E. Barnes

*Structural Mechanics, Second edition*
J. Cain and R. Hulse

*Timber – Structure, Properties, Conversion and Use, Seventh Edition*
H. E. Desch and J. M. Dinwoodie

*Understanding Hydraulics, Second edition*
Les Hamill

*Civil Engineering Materials, Fifth Edition*
Edited by N. Jackson and R. K. Dhir

*Reinforced Concrete Design, Fifth Edition*
W. H. Mosley and J. H. Bungey

*Reinforced Concrete Design to Eurocode 2*
W. H. Mosley, R. Hulse and J. H. Bungey

*Plastic Methods for Steel and Concrete Structures, Second Edition*
S. S. J. Moy

*Highway Traffic Analysis and Design, Third Edition*
R. J. Salter and H. B. Hounsell

*Civil Engineering Quantities, Sixth Edition*
I. H. Seeley

*Understanding Structures, Second Edition*
Derek Seward

*Surveying for Engineers, Third Edition*
J. Uren and W. F. Price

*Fluid Mechanics*
Martin Widden

*Engineering Hydrology, Fifth Edition*
E. M. Wilson

# Civil Engineering Construction Design and Management

## Dene R. Warren
BSc, CEng MICE, MIStructE
*Nene College of Higher Education, Northampton*

palgrave
macmillan

Published by
PALGRAVE

Palgrave Macmillan in the UK is an imprint of Macmillan Publishers Limited,
registered in England, company number 785998, of Houndmills, Basingstoke,
Hampshire RG21 6XS.

Palgrave Macmillan in the US is a division of St. Martin's Press LLC,
175 Fifth Avenue, New York, NY 10010.

Palgrave Macmillan is the global academic imprint of the above companies
and has companies and representatives throughout the world.

Palgrave® and Macmillan® are registered trademarks in the United States,
the United Kingdom, Europe and other countries.

ISBN 10: 0–333–63682–1
ISBN 13: 978–0–333–63682–4

This book is printed on paper suitable for recycling and made from fully
managed and sustained forest sources. Logging, pulping and manufacturing
processes are expected to conform to the environmental regulations of the
country of origin.

A catalogue record for this book is available from the British Library.

Printed and bound in Great Britain by
CPI Antony Rowe , Chippenham and Eastbourne

# Contents

# Preface

This book is written for students on Civil Engineering degree and diploma courses but may also be useful as a reference text for practising Engineers in the design office and on site.

The structure is based on the BTEC unit Civil Engineering Construction A (level H) but its contents have been developed far beyond into practical design and management based on the author's wide experience of industry. The book is a compendium of basic civil engineering construction methods brought together with well known design and management techniques. Wherever possible a summary of design options is presented with case studies and examples to give the reader a feel for the correct design solution. The book is intended to be a gateway to civil engineering, giving the reader an overview of the processes involved and showing the way into many other texts referenced throughout the book.

The book covers a wide range of areas within construction, from management and safety to design and construction, from earthworks and ground water control through to building superstructures and roads. The book begins by reviewing the contractual aspects of pre-construction and construction stages bringing in the options of Client and Engineer choice. Principal contractual and management structures are described and compared, together with the more commonly used financial arrangements. This is developed to illustrate how civil engineering projects can be effectively planned, managed and controlled. Safety is considered in some depth from both the historical viewpoint and in examining today's responsibilities. Fundamental techniques of groundwater control and excavation are reviewed and developed into management and design aspects by the use of case studies and examples. Structures are developed in a similar manner to look at foundations, retaining wall and basement construction and design. Commonly used design techniques are demonstrated with plenty of examples. Different methods of superstructure construction are reviewed and compared giving simple performance and cost comparisons. The stability of structures is also considered with examples. The book concludes with a brief but comprehensive look at drainage, road construction and design.

The author has endeavoured to use up-to-date source material wherever possible. Much information has been gained from current standards, the technical press and direct from industry. Whilst this is the case, however, it must be stated that standards and design methods are frequently updated and the reader must check that the information used from this book is up-to-date at the time.

Dene Warren

# Acknowledgements

I am indebted to all the following people, but most of all to my wife, Linda, who has given me enormous support in the writing of this book. I am very grateful to Roger E Padwick CEng (retired), formerly with Brand Leonard Consulting Engineers of Chelmsford, who has not only taken the time and trouble to proof-read this book to his usual high standard, but who has also been a guiding light and mentor to my career. I have had a great deal of help and assistance from people in the construction industry, too numerous to detail here, but special thanks are given to the following for their kind permission to reproduce information, diagrams and tables from their own publications. Specific items of permission are acknowledged within the text.

- ARC Pipes, Mells Road, Mells, Frome, Somerset BA11 3PD
- Birchwood Concrete Products Ltd, Birchwood Way, Cotes Park Industrial Estate, Somercotes, Derbyshire DE55 4NH
- British Standards Institution. Extracts from BS 8102:1990 are reproduced with the permission of BSI. Complete copies can be obtained by post from BSI Customer Services, 389, Chiswick High Road, London W4 4AL
- British Steel Plc PO Box 1, Scunthorpe South Humberside DN16 1BP
- Fondedile Foundations Ltd, Rigby Lane, off Swallowfield, Hayes Middlesex UB3 1ET
- Health and Safety Executive, Statistical Services Unit, Trinity Road, Bootle, Merseyside, L20 7HE
- HMSO Publications, Room 2b St Crispins, Duke street, Norwich, NR3 1PD
- Keller Ltd, Thorp Arch Trading Estate, Wetherby, West Yorkshire LS23 7BJ
- The Meteorological Office
- Liebherr - Great Britain Ltd, Travellers Lane, Welham Green, Hatfield, Herts
- The Editor, *New Civil Engineer*, Thomas Telford House, 1, Heron Quay, London E14 4JD
- Roger Bullivant Ltd, Walton Road, Drakelow, Burton on Trent, Staffordshire DE15 9UA
- Space Decks Ltd, Chard, Somerset TA20 2AA
- Transport and Road Research Laboratory, Crowthorne, Berkshire.
- Westpile Ltd, Dolphin Bridge House, Rockingham Road, Uxbridge UB8 2UB

# 1 Contract Administration

All construction projects begin with an idea, resulting from a perception of a need, and in the process of its provision aim to make a profit. First we need funding. This may come from a private or governmental source but in either case some form of study is required to prove demand and viability. Professional help in the form of a Civil Engineer or Architect should be employed to help the Client put together a feasibility study and subsequently plan the project.

In this section we will look at the process of administration, how a construction project is taken from an idea, assembled into a contract and then executed. We will overview the contractual arrangements most commonly used in today's industry and also look at some new developments on the horizon. In looking at this process we will examine the following:

- The pre-contract stage, an overview of all activity before construction begins
- The feasibility study – the financial, environmental and engineering assessment of the proposal.
- Gaining authority.
- Choice of the Form of Contract.
- Details of the contract documents.
- Management structure.
- Basis for payment
- Letting the contract

The following terms are used in this chapter:

Client        Initiates the project and is responsible for providing funds for its execution. Sometimes called the Employer or Promoter.

Engineer       Chartered Civil Engineer is appointed by the Client and in a conventional contract is the Client's representative.

Contractor     Organisation responsible for the construction of the works.

## 1.1 PRE-CONTRACT STAGE

When a Client (usually the Government) identifies a need for the construction of civil engineering works a Chartered Civil Engineer is employed to advise him. This is known as the pre-contract stage before work begins on site and before the contracts have been let. The Civil Engineer will investigate the feasibility of the project from the point of view of costs, engineering problems, environmental and statutory requirements i.e. how the project will affect other people. To assist in this endeavour the Civil Engineer may employ a Quantity Surveyor to assess costs and a Solicitor to advise on legal issues if required. It is however common for these professionals to be directly employed by the Client on the recommendation of the Engineer. This working relationship is shown in figure 1.1. The Engineer is employed directly by the Client and acts as an information gatherer from the Quantity Surveyor, Solicitor, Statutory Undertakers such as Gas, Water etc whose apparatus may be affected and Planning Authorities who ensure that the development conforms with adopted planning policies.

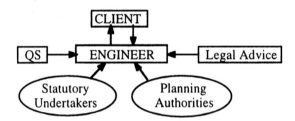

Figure 1.1 Pre-contract stage

On the basis of the information gained by the Engineer's initial investigations and design he will then advise the Client as to the form of construction, the conditions of contract to use and perhaps even the management structure that will best suit the Client's needs. Any special problems identified by the Engineer, such as a high environmental impact due to the presence of contaminated land, will have to be addressed before a detailed design is drawn up and tender prices are sought.

Once the Client and the Engineer have agreed on the construction concepts and expected costs the Engineer will prepare a detailed design in the form of drawings and specification. A Quantity Surveyor will assist by drawing up a Bill of Quantities based on the drawings and specification. The above documents are designed to fully describe the project envisaged by the Client and will form the bulk of the contract documents to be sent out to tender. This is described in greater detail in section 1.5.

To ensure that the Client has the benefit of the lowest costs for the work required the contract documents are sent out to a number of selected tenderers. Three, or sometimes five, tenderers are chosen to submit a price for the work. The returned tenders are thoroughly checked for errors and omissions and the

Contractor who submits the lowest price is usually awarded the contract.

The working relationship described above is a conventional arrangement in that there are many different ways of getting a project started; for example the Client could be an Architect who has already investigated the financial side but who requires help from the Engineer in terms of the environmental impact and construction problems. In that case the Engineer would gather information from site investigations, statutory undertakers, planning authorities and the Architect himself. Alternatively, the Client may be a Contractor who intends to finance and operate the project himself and requires help from an Engineer and an Architect. Whichever is the case, it is vital that all parties work together in an efficient and professional manner to provide the Client with a quality product, at optimum cost, with least risk and a predictable maintenance liability.

## 1.2 FEASIBILITY STUDY

With a project identified, the Client must enlist the assistance of a Civil Engineer or Architect to produce a Feasibility Study. This involves preliminary soil investigation and measurement surveys to investigate technical feasibility. Economics are very important and a Quantity Surveyor is employed to advise on costs. An outline design is put forward by the Engineer to the Client and is costed by the Quantity Surveyor. The cost of the project must fit the Clients budget as well as satisfy his operational needs and several scheme designs may be put forward. When all parties are happy with the proposals only then can detailed design begin and tenders be sought. The Civil Engineer will gather all relevant information on the proposed site, including Local Authority records and Statutory Undertakers. The Engineer will also research the likelihood of gaining permission to carry out the project and may become heavily involved in the public relations exercise. Once this study is complete and an outline design is agreed, a project proposal is put to the necessary authorities for approval.

### 1.2.1 Environmental Impact Studies

As a result of the implementation of European Directive 85/337/EEC a whole tranche of legislative regulations was introduced in 1988/89 to ensure that all projects carry out environmental studies. This is particularly important with road or rail construction, but it is now required that some form of Environmental Impact Assessment (EIA) is made before *any* major project is constructed.

The EIA is a systematic study of the project in the context of the proposed site and provides documentation of the information and estimates of the likely effect of the development on the existing setting. It will contain the following:

- Title, names of parties
- Description and location of the proposed development. Construction activities proposed and programme
- A study of need, planning requirements, alternative constructions methods and details of alternative sites considered

- Scope of the study
- Description of existing environment (called the baseline)
- Identification of key areas of significant environmental impact and
  monitoring programme if necessary. Topic areas will include:
  - Land use, landscaping and aesthetics
  - Geology and topography
  - Hydrology
  - Air quality
  - Wildlife
  - Noise
  - Transport
  - Social consequences
- Predictions of the magnitude of the significance of impact
- Evaluation of the significance of the impacts and any mitigating
  circumstances or measures to be taken to reduce impacts
- Conclusion

The EIA will include a Traffic Impact Assessment (TIA) to examine specific issues of transportation and parking. The Institution of Highways and Transportation publishes guidelines for such a study called Guidelines for Traffic Impact and Assessment[10]. For more information on Environmental Studies the book *Introduction to Environmental Impact Assessment*[3] is recommended.

## 1.3 AUTHORISATION TO BUILD

It is necessary to gain legal authority to carry out the construction of the project. The form of authority necessary depends upon the status of the Client and type of project. For a private sector development, all that is required is legal ownership of the land and Planning Permission, then final design can begin. For Public sector developments such as roads Parliamentary sanction is required. This can be a lengthy process as it is often necessary to carry out a preliminary design and economic appraisal on several alternative routes before public consultation can begin. The consultation process it self can take up to a year and if none of the proposed routes are acceptable then the process must begin again.

### 1.3.1 Planning Permission

This is gained by application to the local Council's planning committee. The committee is composed of local councillors, professionals and lay men and meets once a month, or once every two months, depending upon the population density of the area in question. Notification of the project is posted in public and the committee considers all objections raised by such notification. The committee has the power to grant planning permission which, once granted, is valid for five years. This process takes 3 to 6 months.

### 1.3.2 Parliamentary Sanction

Once the Secretary of State for Transport has decided to go ahead with a project the design has to be largely completed. This is because the site boundary of a road can only be accurately defined once the design is nearing completion and this type of information is needed before it can be considered by the public. An order is published in the local press for six weeks stating the intentions of the Department of Transport. The alternative routes and their environmental assessments are displayed at local venues for public consultation and then four weeks must then elapse so that written objections may be received. (If there are none, the project goes ahead.) When objections are received a minimum of 12 weeks is required to set up a Public Inquiry.

The Inquiry takes one to two weeks and is chaired by an independent Inspector (usually a Civil Engineer). All interested parties will be invited to put their cases to the Inspector. The Inspector's Report, which usually takes two to three months to write, is then sent to the Secretary of State for consideration; the report will also include an Environmental Impact Assessment. If approval is given, it takes four weeks to be confirmed, then six weeks are allowed as a 'Statutory challenge period'. This period is allowed for any persons who may wish to challenge the decision on the grounds of procedural objections. The order is then law and construction may begin. This process takes on average about one year but, if there is fierce opposition, much longer – the Sizewell B inquiry took two years! The Secretary of State may approve, approve with modifications or reject the proposal. Also if the findings of the inquiry are inconclusive or if there are procedural objections a second or even a third inquiry may be ordered. Once the order is made law then the ownership of the land necessary is obtained by 'compulsory purchase order'.

For a private construction project such as a toll road the same procedure of public enquiries has to be undergone. The project is then introduced to the House of Commons as a Private Member's Bill. The Bill is read three times within the House of Commons and then goes to the House of Lords for approval. The Bill then becomes law.

### 1.4 CONDITIONS OF CONTRACT

A simple contract can exist between two parties when one agrees to provide goods or services in return for payment or services provided by the other. Contracts come under the heading of civil law, or the law of tort and in general, English law requires no special formalities for a contract to exist, but for them to be effective they have to be written in a special way. In construction the parties to a contract should define the manner in which a project shall be executed, standards of quality, drawings to be used, time limit, method of payment and the procedure for default in the contract.

There are a number of 'off the shelf' Conditions of Contract available across the construction industry. The traditional form used for Civil Engineering works is the ICE (Institution of Civil Engineers) 6th Edition Conditions of Contract[4].

The Minor Works version[6] introduced in 1988 is a simplified version of the ICE Conditions of Contract. Even more recently the ICE Design and Build Conditions of Contract[7] were published in 1992 and have proved very popular. All of these conditions are written for Civil Engineering and embody the Civil Engineer as the centre of decision making. For Building work the JCT[2] (Joint Construction Tribunal) Form of Contract is most traditionally used; this makes the Architect responsible for decisions.

### 1.4.1 ICE Conditions of Contract 6th Edition 1991

This form of contract is essentially a measurement contract using the Bill of Quantities to emphasise price. It treats design and construction as two separate processes and does not encourage alternative design proposals from the Contractor. It is well know in the industry because until recent times it was the most popular form of contract used, but due to Government pressure for private funding there is a move towards Design and Build. The Department of Transport (DoT) has, however, produced its own conditions of contract (based upon ICE. Conditions) which has some important amendments to the standard terms bringing the Client more into the centre of decision making. These developments will be considered later.

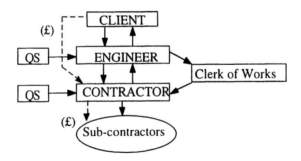

Figure 1.2 Construction stage

There are three main parties to the standard form of contract, the Client, Engineer and Contractor as shown in figure 1.2. In the standard form of contract the Engineer is impartial and will oversee the contract to ensure that the Client's project is built as he expects it to be built. At the same time the Engineer controls the financial payments from the Client to the Contractor. The Contractor has the responsibility to construct the works in accordance with the contract documents which usually include the design of any temporary works necessary for the contract. The quality of the work is overseen by either the Resident Engineer or Assistant Resident Engineer assisted by the Clerk of Works. The Contractor may employ Sub-contractors to carry out some of the work. These Sub-contractors deal only with the Contractor because it is the Contractor who is responsible for the construction of the works and is therefore responsible for the work of the Sub-contractors. The Bill of Quantities is based on an approximate measure at design stage and remeasured at the end of the contract to give the final contract sum. Both the Engineer and the Contractor employ their own Quantity Surveyors to measure work carried out on site. The Quantity Surveyors carry out interim monthly measures upon which interim monthly payments are made; they also carry out the final measure at the end of the contract upon which the final contract sum is based.

The Contractor may claim extra money for extra work or if he has incurred extra costs. He may also claim for extra time as a result of extra work or delays. Under ICE Conditions the Engineer has the power to approve or disapprove such claims on behalf of the Employer and the Employer is then bound by that decision. This has been a source of friction between Client and Engineer in the past, to such an extent that the DoT has written it's own Conditions of Contract which have some important implications. With the new DoT Conditions the Engineer is only empowered to recommend financial decisions to the Client who then makes his own decision.

Under the ICE 6th Edition, the Engineer is responsible for the design and supervision of the works. The Contractor is responsible for construction (including design of temporary works) and programme. The ICE 6th Edition is therefore written for the conventional form of contract and uses the Civil Engineering Standard Method of Measurement 3 (CESMM3).

### 1.4.2 ICE Conditions of Contract for Minor Works 1988

This is designed for contracts priced below £100,000. It is based upon the ICE 6th Edition. Sub-contractors are not allowed unless approved and termination of the contract can only be made under serious circumstances. The wording and documentation for variations and day-work, etc are simpler. The Engineer remains the Client's representative, with powers to make financial decisions on his behalf, and the Contractor remains responsible for construction.

### 1.4.3 ICE Design and Build Conditions of Contract 1992

This document is specially drawn up for the popular 'turnkey' or 'Design and Build' type contracts. Under this type of contract the Contractor will take responsibility for the design and construction and in some cases will operate and maintain the new facility.

### 1.4.4 Fédération Internationale des Ingénieurs Conseils 1977 (FIDIC)

FIDIC[1] Conditions, based on ICE 6th Edition, are sometimes used in overseas civil engineering work. It is usual for Contractors and Consultants to be employed in partnership with similar companies in the host countries and the FICID conditions may then be supplemented by local laws and native contract clauses.

### 1.4.5 JCT Form of Contract

The JCT Form of Contract is used mainly in building construction. It is similar to ICE 6th Edition in that it is a measurement contract with an emphasis on price and treats design and construction as separate processes. In the JCT form the Architect takes the place of the Engineer as the main decision maker. Contrary to the ICE Conditions, the JCT form assumes that the building design is at a more advanced stage of completion before tender prices are sought; this allows the Bill of Quantities to be drawn up in much more detail, giving the Bill of Quantities greater importance in the contract because it can be considered as a 'final measure'. Remeasure is not allowed except in the event of a variation. This is unlike the ICE 6th Edition which accepts that only approximate quantities are known at the design stage and a final measure is made at the end of the contract. The JCT form is also designed to deal with a diverse range of disciplines and skilled tradesmen from steel erectors through to plumbers.

The JCT Form of Contract is in three forms: Standard, Intermediate and Minor Works for large, medium and small contracts. The Standard Method of Measurement for Building Works 7th Edition (SMM7) 1992 is used to compile the Bill of Quantities which must be accurate. Items may not be remeasured at the conclusion of the project. Only variations are remeasured.

### 1.4.6 General Conditions of Contract for Building and Civil Engineering

This form of contract, sometimes called GC/Works[12], originates from the Department of the Enviroment and is a general contract which covers mechanical, electrical, building and civil engineering work. There are different forms of contract within the package: Standard, Minor Works and Lump Sum with or without a Bill of Quantities. The main decision maker is called the Project Manager. The Contractor is encouraged to put forward alternative design proposals but reduction of the contract programme is discouraged.

### 1.4.7 IChemE Conditions of Contract for Process Plant

The IChemE[9] is essentially a cost reimbursement contract which has been used in the water industry for process type projects, where the performance of the completed installation is important and can be tested against well defined parameters. The Contractor undertakes to design and construct all that is necessary to complete the works as described in the Client's specification. The Contractor is encouraged to propose alternative designs and the philosophy is one of partnership between the Client and the Contractor, sharing and solving problems. There are some disadvantages in that it is difficult to compare tenders and the Client will have problems forecasting his financial commitment.

### 1.4.8 Sub-contractors

Within all of the above contracts, Sub-contractors are employed either to carry out a specialist task of which the Main Contractor has no experience or to relieve the Main Contractor of some of the financial risk. There are two types, nominated or domestic. For a nominated Sub-contractor the Client instructs the Main Contractor to enter into a sub-contract with the Client's chosen Sub-contractor. This is used for specialist services such as Mechanical and Electrical plant. Domestic Sub-contractors are employed by the Main Contractor usually on the basis of lowest price tender, but under ICE 6th Edition Conditions of Contract[4] the Main Contractor may only enter into such a sub-contract with the Engineer's agreement.

## 1.5 CONTRACT DOCUMENTS

This section discusses the contents of the contract documents and gives an indication as to the function of each. Taking as an example a standard measurement contract with quantities the contract will consist of the following documents:

- Instructions to Tenderers
- Form of Tender
- Conditions of Contract
- Specification
- Bill of Quantities
- Drawings
- Form of Agreement
- Performance Bond

Each of these documents are interdependent upon each other to define the contract. For example, the Specification will be drawn up with the Drawings to define in words, and in greater detail, that which is shown on the Drawings. Items on the Drawings will reflect the Conditions of Contract and the Bill of Quantities will be drawn up from information provided by the Drawings, Specification, Conditions of Contract, etc.

### 1.5.1 Instructions to Tenderers

These instructions are intended to assist the Contractor in the compilation of his tender so that it is presented in the form required by the Client and Engineer. It consists of:

a) All documents to be submitted with tenders such as
   i) Programme
   ii) Method of Work Statement
   iii) Quality Statement
   iv) Declaration of the worthy financial and technical status
b) Place, date and time for the delivery of tenders.
c) Instructions on the visiting of the site.
d) Instructions on whether tenders based on alternative designs will be

considered and if so under what conditions they may be submitted.
e) Notes, drawing attention to any special Conditions of Contract, materials and Methods of Construction and unusual site conditions.
f) Instructions on the completion of the Bill of Quantities, Schedule of Rates and the Production of Performance Bonds.

In the case of Invited Tenders a) iv) is not required, since the tenderer is selected from a list of approved tenderers who have already been vetted in this way. Open tendering is when tenders are considered from any Contractor who wishes to make a bid following advertisement in the technical press.

### 1.5.2 Form of Tender

This is the Contractor's written offer to execute the work in accordance with the contract documents and states the following:

- Total tender sum
- Time for completion
- Any other special offers such as variation to the Specification if some advantage may be gained.

A Model Form of Tender is published in the Appendix to the ICE Conditions of Contract 6th Edition.

### 1.5.3 Conditions of Contract

These have been considered in some detail in section 1.4.

### 1.5.4 Specification (Performance Specification or Method Specification)

A Specification describes in words the work to be executed in detail. The detail is such that it will describe, for instance, the type of paint to be used and the dimensional tolerances within which the works will be constructed. There are two types of specification used: Performance and Method Specification.

*Performance Specification*

Here, details are given of the character and quality of the materials to be used, the standard of workmanship, any special responsibilities of the Contractor not included in the Conditions of Contract and also the facilities available to the Contractor. The Specification should require the Contractor to submit a Programme (Contract Programme against which delays or otherwise are defined), Methods of Works, Quality Statement and details of temporary works, all for the approval of the Engineer. Care must be taken not to contradict the Conditions of Contract. The Specification makes reference to all other information relevant to the contract, i.e:

- All other contract documents.
- Soil investigations and reports.
- Statutory and legal restraints special to the site, i.e. underground services, rights of way, etc.
- Safety and Security.

*Method Specification*

The Specifications may describe the method of working to the whole contract or to a portion of the work. This means for example that, instead of stating the strength of the concrete required, it states the proportion of materials and the method of mixing in the hope that the strength will be achieved. This has the obvious disadvantage that the responsibility for achieving a stated strength is that of the specifying Engineer and not the Contractor, provided that the method is followed precisely as laid down in the Specification. This form of Specification is rarely used these days, since it will restrict the Contractor's options and almost certainly cost more money.

The Specification can be a substantial document which must be writen with care and it is often eaisier to use an existing specification and adapt it for the project in question. Standard Specifications are published by interested parties such as the Specification for Highway Works published by the Department of Transport and the National Building Specification.

### 1.5.5 Bill of Quantities

The Bill of Quantities is composed of brief descriptions of the items of work to be undertaken within the contract, against which is written an approximate quantity. This is drawn up by the Quantity Surveyor for the Engineer or Architect and will become part of the contract documents. The Bill of Quantities is prepared in accordance with the Civil Engineering Standard Method of Measurement 3[5] (CESMM3), for civil engineering work and Standard Method of Measurement 7th Edition[11] (SMM7), for building works.

When tendering, the Bill of Quantities is sent out as part of the package of contract documents for the Contractor to price. The Contractor enters a unit price or rate against each item of work, the quantities are then multiplied by the rate and the price for the item of work is established; all the item prices are then added up to give the contract price. Provision may be made for variation and additional work within the Bill of Quantities by including provisional sums.

Setting-out items of works in such a formal way is advantageous both to the Client and the Contractor because it gives a means by which tenders may be compared and a basis upon which variations of work may be priced. When the tender is first submitted to the Client, the Engineer can check the rates and prices to ensure that they are mathematically correct and that the appropriate rates are correct. Tender submissions are often completed to a tight time limit and it is not uncommon for mistakes of this nature to be made. The Contractor may inadvertently quote a low rate for an item of work which could result in financial hardship or even jeopardise the project. The Engineer should also look at the effect of increased quantities in particularly 'risky' items of work such as earthworks or piling and compare rates submitted by the various tenderers. As the project continues, progress payments can be quantified by looking at specific work items which have been completed; and at the end of the contract the actual quantity executed under each item of work is remeasured and valued at the quoted rate, this is called the final account. In these days of fixed price and design and build contracts it can be tempting to dispense with a Bill of Quantities but that must be considered as a false economy in the face of variations, claims and ever tighter profit margins.

### 1.5.6 Drawings

The drawings should give details of all the work to be undertaken in the contract, including existing site topography and services. Specifications and drawings take precedence in the event of errors or disputes.

### 1.5.7 Form of Agreement

This is a legal undertaking entered into between the Client (Promoter) and the Contractor so that the work will be carried out in accordance with the contract documents. A model Form of Agreement is published in ICE Conditions of Contract 6th Edition[4] Appendix and should be signed and sealed by the Client and Contractor. The completion of the document is usually delayed by other events, but a safeguard is built into the Form of Tender which normally states that the receipt by the Contractor of a Letter of Acceptance from the Client will constitute a binding contract.

### 1.5.8 Performance Bond

This is a document which states that a specified bank or insurance company will pay a specified sum to the Client if the Contractor fails to discharge satisfactorily his obligations under the Contract. A Model Form of Bond is published in the ICE Conditions of Contract 6th Edition[4].

## 1.6 MANAGEMENT STRUCTURES

Once the Client has the necessary authority for the project, he then has to decide under which type of contract the project should be run. Again the Engineer will advise the Client regarding the main options, which are:

- Conventional Contract
- Management Contract
- Construction Management
- Design and Build (Turnkey) Contract

### 1.6.1 Conventional Contract

In a Conventional Contract the Client appoints a Civil Engineer to design and supervise the construction of his project. The detail of this type of contract has been considered previously, see section 1.4.1. The Engineer is the Client's representative on site and may make financial decisions on his behalf. This has recently been modified in the DoT's own form of contract so that the Engineer can only recommend financial decisions to the Client. The Engineer invites tenders and appoints the lowest tender on the Client's behalf. The Engineer recommends the appointment of a Quantity Surveyor who then advises the Engineer on variations and quantities.

### 1.6.2 Management Contractor

Here the Client appoints the Civil Engineer to carry out the design but also appoints a Contractor to manage the construction simultaneously. So it is the Management Contractor who goes out to tender and is the Client's representative on site. This is shown in figure 1.3.

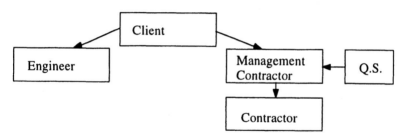

Figure 1.3 Management contract

There are a number of advantages to be gained from this arrangement over the conventional contract. The first is that because there is a division of activities between Engineer and Management Contractor there is a potential for saving time by overlapping the design and construction process. The second advantage is that there can be a greater flexibility of design using the Management Contractor's expertise. Alternative designs can be proposed by the Management Contractor which must then be evaluated in detail by the Engineer. Another potential advantage is a possible improvement in cost forecasting based upon the Management Contractor's knowledge of his own resources and experience. A disadvantage from the Client's point of view is a potential increase in administration costs.

### 1.6.3 Construction Management

This is when the Contractor is employed directly by the Client. The Client then either has his own representative on site or the Management Contractor acts as 'agent to the Client'. This is shown in figure 1.4.

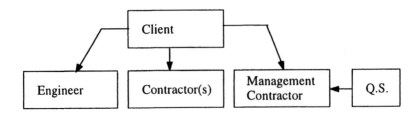

Figure 1.4 Construction management

This type of management structure has the same advantages as for the Management Contractor system but also gives the Client much greater control of work on site. Such a system is particularly attractive to a Client who has the expertise within his staff to manage the situation and has his own special requirements to fulfil on site. The disadvantages are that with increased control the Client will also have increased exposure to risk of extra costs on site perhaps due to unforeseen circumstances. Another disadvantage is of course an increase in administration costs.

### 1.6.4 Design and Build (Turnkey) Contract

With Design and Build Contracts the Client employs an Engineer to produce an outline design. The Client then asks a number of Design and Build Contractors to price the scheme. This procedure is shown in figure 1.5.

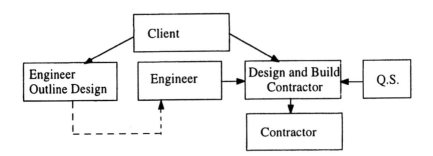

Figure 1.5 Design and Build

Each Design and Build Contractor will employ his own Engineer and Quantity Surveyor to assist and submit a tender price by the given dead-line. The Client generally employs the Design and Build Contractor who has submitted the lowest bid. The successful tenderer is then given overall responsibility to both design and build the project to the Client's Specification. The Engineer initially employed by the Client is usually employed by the successful tenderer to draw up the detailed design.

This type of contract is so popular that the ICE has its own Conditions of Contract Design and Construct[7]. The advantage of this system is that once the initial Specification is drawn up at the beginning of the process the Client will have very little involvement until the project is complete, except for staged payments which require some sort of check on percentage completion. This type of contract is usually run on a fixed price basis so that the Client will know early on in the contract exactly what his financial commitment is. The disadvantage is that there is no independent supervisor on site to check quality which may suffer under tight deadlines and budget. For this reason it is common for the Client to employ an independent site supervision Engineer who reports directly to the Client on a regular basis.

## 1.7 BASIS FOR PAYMENT

Any of the preceding conditions of contract may be used with different methods of payment. For example, the Contractor may be paid a lump sum at the end of the contract or may be paid in monthly interim payments. The following is an overview of typical methods of payment in the construction industry.

### 1.7.1 Bill of Quantitiesfor Measurement Contracts

Measurement contracts are where the cost of the contract is based upon a pre-measured Bill of Quantities taken from the Engineering Drawings and Specification such as in the ICE Conditions of Contract or the JCT Form of Contract. The Bill of Quantities is looked at in more detail in section 1.5.5.

### 1.7.2 Schedule of Rates

The Schedule of Rates is a comprehensive list of items of work covering the operations that the Client may require. No quantities are given and the Contractor will price work per square metre, or per cubic metre in the case of concrete. The Contractor is then expected to carry out the operations for the price quoted no matter what quantity of work is required. Here the Client can only guess at the final cost, but it is used for small repetitive operations or work that must be carried out in a hurry.

### 1.7.3 Lump Sum Contracts

This is when the contract price is a fixed lump sum at the end of the job. The Contractor undertakes to carry out the work as shown on the Drawings and detailed in the Specification for that fixed sum. Sometimes a Schedule of Rates is linked to this type of contract in case there are any alterations. The advantage to the Client is that if no alterations are made then he knows exactly how much his project will cost; which is useful when there is little likelihood of variations.

### 1.7.4  Cost Reimbursement Contracts

This is where all the expenditure of the Contractor is paid for by the Client plus a fee. The fee is for the use of the general resources of the Contractor's organisation known as 'overheads', plus a small profit.

This type of contract has four variations:

- Cost and Percentage Fee Contract: Here the fee is a fixed percentage of the allowable cost.
- Cost and Fixed Fee Contract: The fee is fixed based upon an agreed estimate of cost.
- Cost and Fluctuating Fee Contract: The fee is tied to project cost by some form of fee formula.
- Target Contract: The fee is tied to the actual cost of the work. The actual cost is compared to an estimated cost and if money is saved the fee goes up.

### 1.7.5  Design, Build, Finance and Operate Contracts

Here the Client states, through his Engineer, his requirements in broad and general terms with an Outline Specification. Sometimes the Specification can be very detailed, but in all cases the Contractor is given the responsibility for the providing funds, design, construction, maintenance and operation of the Client's project. The Contractor may be asked to operate the facility for a set period of time during which he aims to recoup his funds with profit. After the fixed time the project will pass into the hands of the Client. Such contracts can be lucrative but the financial risk for the Contractor may be large. To tackle projects in this way a number of Contractors may form themselves into Consortia for joint venture projects. Typical examples of such joint ventures are the Channel Tunnel, completed in 1994, and the Dartford River Crossing, completed in 1991. In the case of the Dartford River Crossing the Consortia agreed to construct the bridge and operate it and the adjacent tunnels for a fixed time and then hand the project back to the Client free of charge. In the meantime motorists pay toll charges to pay for the project, financing cost and profit.

### 1.7.6  Lane Rental

This contract is based on the conventional ICE 6th Edition Contract with a Bill of Quantities. Included in the contract is an allowance for the Contractor to rent each lane of the road at a fixed price for a fixed term, say £10,000 per week per lane for 40 weeks. As the contract proceeds the Contractor pays the Client rent in accordance with the amount of time each lane has been coned off. If the Contractor completes before the fixed term, say at 38 weeks, he may keep the balance of the rental money yet to be paid to the Client. If the Contractor completes after the fixed term, say at 42 weeks, the Contractor must pay the extra rent. In desperate circumstances a sizeable bonus may be offered to the

Contractor to complete early in addition to lane rental payments. One can imagine the stress under which the Contractor's staff operate.

Since roadwork delays have been high on the political agenda recently, the Contractor's predicted time for completion can play a major factor in awarding the contract. The precise amount charged as rent is calculated using the Quadro computer programe which calculates the cost incurred by motorists due to queues and delays.

## 1.8 LETTING THE CONTRACT

Once the Client has considered all the options, gained authority to go ahead and obtained the finances the time comes to let the contract. This is usually done on a tender basis to ensure maximum competition and so the lowest cost to the Client. There is considerable debate as to whether acceptance of the lowest price is in fact in the Client's best interest because low prices may mean poor quality and that leads to higher maintenance cost throughout the product's life. This is where the Engineer's work begins to be tested to the full because if the Drawings and Specification are correct the Client should expect to be able to drive the Contractor down to the lowest price and still get the quality he asked for.

Tender documents are sent out to 3 or 4 tenderers selected from a list of approved Contractors. Each Contractor is given the same length of time to draw up an estimate of the cost of the work; this is called the 'Tender Sum'. Each tenderer will look at all of the contract documents in great detail and carryout a complete appraisal of the proposed works in terms of buildability and costs. In an effort to gain an edge on the competition a contractor may propose an alternative design, but he does so at his own risk because it may not be accepted by the Client or Engineer. All tender bids are opened together, usually under the supervision of an independent scrutineer, and the information is given to the Client for his decision on who to appoint. The Engineer is sometimes required to examine each bid and present a tender report to the Client together with recommendations. Within this report the Engineer should check that there are no obvious errors within the calculations used to work out the tender sum. He should check that funding sources are secured and that the contractors have the resources available to commence construction when required. The Engineer should also check that access is readily available. To let a contract with problems in any of the above areas will lead to claims at a later stage. Obviously, this is an anxious time for the tenderers and the Engineer and Client should endeavour to make their decision as quickly as possible. The Client may not accept any of the bids if they are all far above his own estimates and can ask tenderers to resubmit their bids. This costs money and the Client may have to reimburse the tenderers for the work. Some Clients may wish to enter into post-tender negotiations with the lowest two bids, but this can put undue pressure on the tenderers and should not be encouraged.

## 1.9 THE FUTURE

In 1993 the Institution of Civil Engineers published a new Form of Contract called the New Engineering Contract[8] (NEC). This Form of Contract is different from previous forms because it has been written in plain English with the minimum of legal terms. It has simple flowcharts to aid understanding of contract procedure and embodies a change from the traditional 'claims' culture to a more conciliatory approach. The NEC aims to eliminate adversarial relations by removing the traditional barriers between Client, Contractor and Engineer. Each party to the contract is asked to substitute individual interests for the common goal of completion of the contact. This is called the 'partnering' concept and aims to promote closer working relations. For example, it allows staff of the Client and Contractor to work together in areas such as design, administration and management, in the same offices if necessary. This should prevent duplication of effort, allow pooling of resources and promote sharing of ideas.

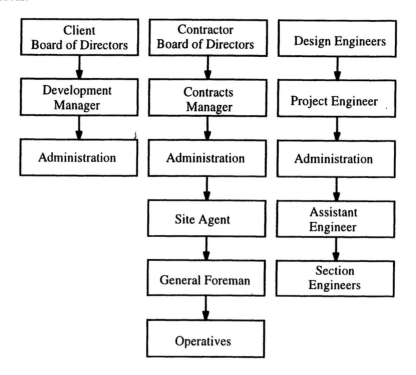

Figure 1.6 Traditional set-up
*Diagram is based on information courtesy of The New Civil Engineer*

The Government appointed Sir Michael Latham to carrying out a review of the construction industry with a view to the standardardisation of working practices and improvement of efficiency. The Latham report, published in

August 1994, recommends that a revised NEC form of contact (probably to be called the New Engineering and Construction Contract) is adopted across the Construction Industry and this is likely to be made law in 1996/97. On the ground, the system works as shown in figure 1.7 as opposed to the traditional approach shown in figure 1.6. The main change for the Engineer is that his role as defined in the ICE 6th Edition has been divided up into the Designer, Project Manager, Supervisor and Adjudicator.

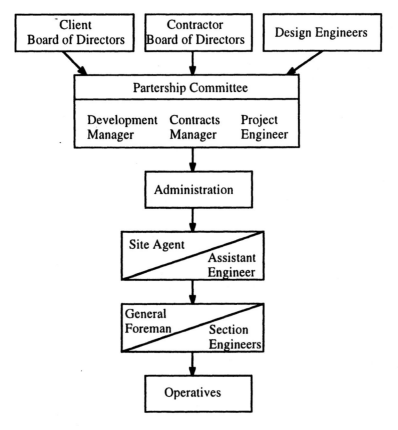

Figure 1.7 Partnering set-up
*Diagram is based on information courtesy of The New Civil Engineer*

For further reading on Contract Administration, see *Civil Engineering Contract Administration and Control* by I Seeley[13].

## 1.10 REFERENCES

1.  Fédération Internationale des Ingénieurs Conseils and the Fédération Internationale Européenne de la Construction. *The Conditions of Contract*

*(International) for Works of Civil Engineering Construction, 1987 (FIDIC)*,
Thomas Telford. ISBN 0-7277-0127-4

2.  R. F. Fellows. *JCT Standard Form of Building Contract, a commentary for
    students and practitioners*, 3rd Edition,1995, Macmillan Press ISBN 0-333-
    64624-X
3.  J. Glasson, R. Therivel and A. Chadwick. *Introduction to Environmental
    Impact Assessment*, 1994, University College London Press ISBN 1-85728-
    118-7
4.  Institution of Civil Engineers, Association of Consulting Engineers and
    Federation of Civil Engineering Contractors. *ICE Conditions of Contract,
    6th Edition, 1991, Conditions of Contract and Forms of Tender, Agreement
    and Bond for use in connection with works of Civil Engineering
    Construction.* Thomas Telford. ISBN 0-7277-1617-4
5.  Institution of Civil Engineers and Federation of Civil Engineering
    Contractors. *Civil Engineering Standard Method of Measurement 3*, Third
    Edition 1991 (CESMM3). Thomas Telford. ISBN 0-7277-1561-5
6.  Institution of Civil Engineers, Association of Consulting Engineers and
    Federation of Civil Engineering Contractors. *ICE Conditions of Contract
    for Minor Works 1988.* Thomas Telford. ISBN 0-7277-1329-9
7.  Institution of Civil Engineers, Association of Consulting Engineers and
    Federation of Civil Engineering Contractors. *ICE Design and Build
    Conditions of Contract 1992 Conditions of Contract and Forms of Tender,
    Agreement and Bond for use in connection with works of Civil Engineering
    Construction*, Thomas Telford. ISBN 0-7277-1695-6
8.  Institution of Civil Engineers. *New Engineering Contract*, Thomas Telford,
    1993, ISBN 0-7277-1664-6
9.  Institution of Chemical Engineers. *IChemE model form of conditions of
    contract for process plant, Reimbursable contracts*, 2nd Edition, IChemE,
    Rugby, 1992.
10. Institution of Highways and Transportation, *Guidelines for Traffic Impact
    and Assessment.* , 1994, IHT, ISBN 0-902-933-12-4
11. Royal Institution of Chartered Surveyors and the Building Employers
    Confederation. *The Standard Method of Measurement for Building Works
    7th Edition (SMM7)*, 1992, The Royal Institution of Chartered Surveyors
    and the Building Employers Confederation, ISBN 0-85406-360-9
12. Royal Institution of Chartered Surveyors. *General Conditions of Contract
    for Building and Civil Engineering Works* (GC/Works/1), Edition 3, 1990,
    12/90. RICS.
13. Ivor H Seeley. *Civil Engineering Contract Administration and Control,
    Second Edition*, Macmillan Press, 1993. ISBN 0-333-59743-5

# 2 Control

Once an idea for a project has crystallised into a detailed design and appropriate contractual arrangements have been made, work can begin on site. Much effort has already been expended by the Client and his advisors to define what is required and it is now the duty of the supervising Engineer and the Contractor to ensure that the Client is satisfied with the final product. That will mean a project which is completed on time, within budget and to the quality expected by the Client. In the previous chapter we looked in some detail at contractual administration of construction; in this chapter we shall look at the management structures which exist on site and how they reflect contractual arrangements. We begin by looking at the staffing structure for Client and Contractor and go on to examine the details of financial control on site. An overview of quality control is followed by a detailed look at the techniques available to site managers to exercise operational control.

## 2.1 CLIENT CONTROL

The amount of control the Client wishes to exercise will be reflected in the type of contract chosen for the job. With a Design and Build or a lump sum contract, for example, the Client has almost no control over the activities of the Contractor. Once on site the Client carries little risk of increased costs. On the other hand in a Cost Plus contract the Client has absolute control but is exposed to greater financial risk. The more control exercised by the Client, the more risk he exposes himself to. In a Cost plus contract the Client is given all details of the Contractor's activities and cash situation, issues instructions to the Contractor on a regular basis and acts more or less as a Project Manager. The Client may be liable for extra costs; for example, if particularly bad ground conditions are found then the Client must tell the Contractor what to do to get over the problem (advised by a Chartered Engineer) and the Contractor will require payment for this extra work. On the other hand if the contract was a lump sum the Client may avoid extra expenditure because the Contractor priced the job at his own risk.

In a conventional Contract, ICE 6th edition[3], the Client's control is delegated to the Resident Engineer (RE). The Engineer must remain impartial in the execution of his professional duties, ensure that the contract is completed and fulfil his legal obligations of care. The RE is aware of the Contractor's need to make a profit but will enforce the terms of the contract agreed by all parties to ensure that the Client gets a fair deal. To discharge this responsibility the RE will assemble a team of Engineers and experts, shown in figure 2.1, who have experience in the field of site supervision.

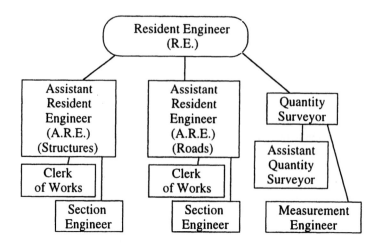

Figure 2.1 Client's representatives on site

The RE's team will monitor progress and quality during construction. Every month they will recommend that a payment is made to the Contractor by the Client which reflects progress; these are called monthly interim payments. Problems will come up from time to time which may need extra work from the Contractor and/or design work by the Engineer. The payment for such work is agreed with the Contractor by the RE on behalf of the Client.

## 2.2 CONTRACTOR CONTROL

The Contractor wishes the Client to be satisfied with the finished product so that he will be asked to tender for the Client's next project, but he is also interested in making a profit. This means that he will endeavour to complete the contract as quickly as possible, in accordance with the drawings and quality defined in the specification, thus saving time and money. There will inevitably be situations that arise which were not foreseen at the design and planning stage and the Contractor will want to ensure that these situations will not delay him and that a fair price is agreed if extra work is needed. To achieve these aims the Contractor will employ an experienced management team whose job it is to maintain control of the works. A typical team set-up is shown in figure 2.2.

Control is achieved by experienced professionals keeping a close watch on operational activities, cash flow in relation to progress and the quality of the work carried out in relation to the job specification. As you would expect this is a complex task and a management team such as that shown in figure 2.2 would be used for a medium sized highway contract of approximate value £2m to £7m (at 1995 prices).

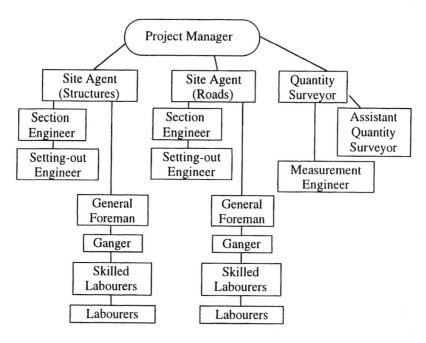

Figure 2.2 Contractor's management team on site

## 2.3 FINANCIAL CONTROL

Financial control is as important as quality control because for the Contractor one depends upon the other. There are a number of tools which are available to the construction management team that help with control. The first and most important is the programme of operations which is looked at in some detail in section 2.6. Once a programme is established the resource requirements can be drawn up and from these cash flow estimated via the cost/value curve.

### 2.3.1 Basic Cost/Value Curve (S-curve)

The cost value/curve is a graph which can be drawn showing the cumulative amount of planned resources to be provided by the Client and the Contractor against time. This allows both the Client and the Contractor to estimate cash flow requirements; an example is shown in figure 2.3. Here, Curve 1 is the planned cumulative value of work in accordance with the contract. Curve 2 is the planned payments to be made by the Client to the Contractor in accordance with the programme of works and the Bill of Quantities. Curve 3 is the actual value of work completed to date in accordance with interim payments. The Contractor can also draw up S-curves of actual costs to date against planned costs and so on. The area of the graph between Curves 2 and 3 indicates the amount of money required by the Contractor to 'finance' the project. This is needed because there will inevitably be a delay between construction of the work and payment. This is known as the 'Capital Lockup' and can be as much as £0.5m for a medium sized contract. To overcome this problem Contractors sometimes use 'front end loading' to reduce the capital lockup. This is where items of work to be carried out at the beginning of the contract are adjusted to increase their value relative to items of work to be carried out in the middle of the contract. This allows the Contractor to gain more money at the beginning of the contract sufficient to fund the capital lock up without affecting the overall contract value. Clients are uncomfortable with this tactic and it must be checked for by the Engineer at the tender stage.

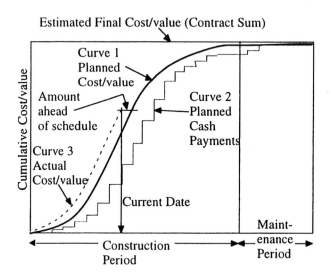

Figure 2.3 The S-curve

### 2.3.2 Documentation

During the contract if all went according to plan, and there were no unforeseen problems, the work would proceed in accordance with the programme and interim payments would be made in accordance with the cost/value curve. There will, however, inevitably, be problems that arise which will require instruction from the Engineer and may cause the Contractor extra costs. The following documentation is typical of that needed to maintain control of these costs.

*Confirmation of Verbal Instruction Sheets (CVI),*

The Contractor must ensure that all instructions are recorded on a CVI within 24 hours and countersigned by the member of the RE's staff who issued the instruction. CVI's are not binding in the contract and must be agreed and confirmed by letter between the parties. An example of a CVI is shown in figure

2.4. If the instruction involves work extra to the contract then the work must be measured and agreed between parties on a Measurement sheet, an example of which is shown in figure 2.5. This is a situation where trust and cooperation are needed between the Contractor's and Engineer's staff. If the Contractor had to wait for written confirmation of an instruction before acting expensive delays can ensue, and so it is imperative that the RE's staff have the authority to make decisions and that the Contractor knows he will receive fair recompense for any actions taken as a result.

*Record of Measurement*

Record of Measurement sheets are often used to record measure and agree items in the Bill of Quantities that could not be measured at the design stage or that may not be accessible to measure at the end of the contract. For example the quantity of soft ground unsuitable to support a road cannot be measured until it is excavated and will not be accessible when the road is complete. The quantity is measured and agreed by site staff of the Engineer and Contractor and recorded on the Measurement sheet shown in figure 2.5.

*Variation Order (VO)*

A Contractor will not be paid for any work over and above that defined in the contract unless it is confirmed by a Variation Order. This is the official notification to the Contractor that the contract has been varied. It is written by the Engineer on behalf of the Client which means that the Client will pay extra money. All instructions and measurement sheets must be converted to a VO, as shown in figure 2.6, before the Contractor can expect payment.

*Daywork Sheets*

When extra work is proposed which is not priced in the Bill of Quantities it can be priced in three ways:

- by an agreed lump sum
- using existing rates in the Bill of Quantities
- using Daywork rates.

Daywork rates for plant, labour and materials are agreed within the Bill of Quantities at the tender stage. The Contractor records the amount of Labour, Plant and Materials used each day in carrying out activities directed by the RE on behalf of the Client. Daywork is rarely agreed to by the RE because it is an expensive way to get work done and requires close supervision by the Client. It is only agreed to if the work to be carried out is not comparable to any existing item in the Bill of Quantities and cannot be predefined to price as a lump sum. Daywork can only commence once a Daywork Order has been signed by both parties. All work carried out under the Order must be recorded daily on Daywork Sheets. Again, these are to be agreed and signed by both parties.

*Site Diary*

All site staff are responsible for keeping a record of site activities in their own diaries on a daily basis. These records are usually the property of their employers. An official Site Diary is kept by a senior member of staff for both Contractor and Engineer. These can serve as valuable sources of information in the event of claims, disputes or accidents.

*Progress Meetings*

Progress meetings are held on a monthly basis, they are usually chaired by the RE and attended by senior staff of the Client, Engineer, Quantity Surveyor and Contractor. All aspects of the contract are discussed within an agreed agenda which includes, a progress report from the Contractor indicating sources of delay if any, a safety report from the Planning Supervisor (see section 3.4.3), outstanding information or instructions required from the RE and an indication from the Contractor of the programme of work in the coming weeks. Accurate minutes are kept for future reference.

## 2.4 QUALITY CONTROL

Quality control includes both testing materials and inspecting work to ensure that they conform to the required standard. The standard of quality to be achieved is laid down in the Contract Specification which as we have seen earlier is a contract document which refers to the appropriate British Standards and Codes of Practice for Construction. The Contractor is under a contractual obligation to comply with the defined quality standards. This is usually supervised by the Resident Engineer represented by the Clerk of Works or a the Building Inspector who is an independent third party employed by the local council. Both have the power to stop work if the required standards are not being met.

## 2.5 QUALITY ASSURANCE

This is the term given to a comprehensive monitoring systems of report and record sheets verifying application of quality control. The process is defined in BS EN ISO 9000[1] and any company wishing to become recognised for quality assurance must be registered with a certification body. The British Standards Institute (BSI) is a certification body, but there are also other independent bodies. The BSI is an impartial independent organisation which would vet any applicant company to ensure that quality assurance procedures that are in place conform to the guidance given in BS EN ISO 9000[1]. There are three grades of accreditation:

**Bloggs Construction Co Ltd**

**Nene College, St Georges Avenue, Northampton**

Sheet Number 0001

To:

Date

From:

Site:

Confirmation of verbal instruction:

Signed.......................Client's representative          Date:

Signed.......................Contractor's representative

The above instruction is to be considered to be a Variation Order if not confirmed within seven days of the above date.

Figure 2.4 CVI form

Sheet Number 0001

# Bloggs Construction Co Ltd
**Nene College, St Georges Avenue, Northampton**

From Resident Engineer:

To Contractor:

### MEASUREMENT SHEET

Contract Reference........................          Location......................

Description of measured works with reference to Bill of Quantities

Sketch of measure:

Signed........................Client's representative          Date:

Signed........................Contractor's representative

Figure 2.5 Measurement sheet

Sheet Number 0001

# Bloggs Construction Co Ltd
**Nene College, St Georges Avenue, Northampton**

From Resident Engineer:

To Contractor:

### VARIATION ORDER

Contract Reference
Description of variation with reference to Contract Documents

_____
_____
_____
_____
_____
_____
_____
_____
_____
_____
_____
_____

**Suggested Method of Payment**

_____
_____
_____

_____

Signed........................Client's representative        Date:

Signed........................Contractor's representative

Figure 2.6 Variation order

| First party assessment | for suppliers of materials |
| Second party assessment | for Contractors and installers |
| Third party assessment | for Designers |

These grades are not a matter of increasing quality but are simply different levels of accreditation for different levels of complexity and applications.

An example of quality assurance in action would be that of a drawing office which receives drawings from a Client requesting a design. The Engineer in charge would probably telephone the Client to say thank you and ask when the design was wanted by etc, and this may be confirmed by letter. In a quality assurance system the drawings would be stamped with the date received and a register drawn up listing each drawing received. A copy of this 'drawing received register' would be sent to the Client together with the Engineer's thank you letter so that the Client then has a record of the drawings he sent to the Engineer. The drawing office also has a record of the drawings received so that if at some point in the design a drawing is lost the Engineer can trace which one it is. A job sheet is drawn up which records the facts of the design, i.e. Client's name and address, drawings received, name of the Engineer in charge, dead line for completion, etc. We can see then that the office activity is recorded in a standard way, providing evidence by filling in the appropriate forms.

The important thing about using a Quality Assurance system is that it provides a 'quality trail' which can be checked to verify that procedures are followed. Thus with the drawing register example, it is important that changes and amendments are accurately logged. There may be several derivatives of the same drawing, and it is necessary to know a) if they have been checked and b) where and when they went. Quality Assurance is therefore a defined system which records the application of office activities to a set standard.

The defined system of set standards are laid down in a Quality Assurance Manual (drawn up by the drawing office itself) and this manual will define exactly what the procedures are and the records which must be kept for different circumstances. Obviously every circumstance cannot be predicted and so the procedures and forms are designed to be flexible. For the above example the manual would state that a drawing received register must be drawn up, an acknowledgement letter must be sent and a job file raised with a job sheet at the front. The manual may have in it the design of the drawing register and job sheets to be used.

The processes defined in the Quality Assurance Manual are checked by a Quality Assurance Audit. Every three or six months an internal audit is carried out by a member of the company who will check that the procedures defined in the manual are indeed being carried out and counsel the people involved if not. Every year an external audit is carried out by an independent inspector. The inspector will check that the standards as laid down in the quality assurance manual are being followed and he has the power to withdraw registration if they are not. The firm may be advised to make improvements and a re-inspection organised in say three months' time.

There are many advantages to being diligent with record keeping not least as being seen by Clients to have an efficient and professional method of working, but these systems are costly. There is a great deal of resentment in design offices about the seemingly needless waste of effort required by these proceedures. For this reason it is important to build the QA system around existing office procedures and to ensure that staff are fully involved in its design and operation. The important issue behind the Quality Assurance system is that all employees realise that everybody is responsible for the quality of their work and that there is much to be gained from getting it right first time. Quality is the responsibility of *all* operatives. For further information see *Quality Assurance in Building* by Alan Griffith[2].

## 2.6 PROGRAMMING

The programming of construction activities is the key to efficient and profitable contracts. It is the main technique used by the construction team to maintain control of the works, but it is often described as a 'wish and a prayer' because of its speculative nature. Computers can be used to programme complicated sequences of work and in the hands of an experienced member of staff can become a powerful tool. This section reviews a number of programming methods; including Bar Charts, Linear Programmes, Line of Balance and Network Programmes. All these methods map activity as a function of time.

### 2.6.1 Bar Chart

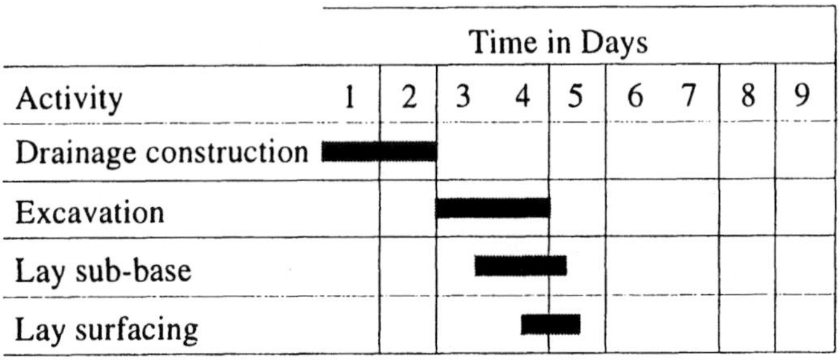

| | Time in Days | | | | | | | | |
|---|---|---|---|---|---|---|---|---|---|
| Activity | 1 | 2 | 3 | 4 | 5 | 6 | 7 | 8 | 9 |
| Drainage construction | ▬▬▬ | | | | | | | | |
| Excavation | | | ▬▬▬ | | | | | | |
| Lay sub-base | | | ▬▬ | | | | | | |
| Lay surfacing | | | | ▬▬ | | | | | |

Figure 2.7 Bar chart

This is the easiest method of planning and programming works and is most commonly used. Activities are listed on the left hand side. The time scale is drawn horizontally and bars drawn on the chart represent the time when work will proceed on each activity. The example shown in figure 2.7 looks at the construction of a section of road over a nine day period. Although the units for time are days this type of programme could equally have used units of weeks or months. There are one or two refinements to this type of programme that can be used to give more information, the Linked Bar Chart and the Progress Bar Chart.

### 2.6.2 Linked Bar Chart

The Linked Bar Chart is based upon the simple bar chart but includes lines across activities which cannot, by their nature overlap. For example, in figure 2.8 the excavation of formation cannot begin until the drainage is complete and so a line is drawn across the ends of these activities. Laying the sub-base, however, can commence after only half a day of excavation and so an arrow is drawn to indicate the earliest start time.

| | Time in Days | | | | | | | | |
|---|---|---|---|---|---|---|---|---|---|
| Activity | 1 | 2 | 3 | 4 | 5 | 6 | 7 | 8 | 9 |
| Drainage construction | | | | | | | | | |
| Excavation | | | | | | | | | |
| Lay sub-base | | | | | | | | | |
| Lay surfacing | | | | | | | | | |

Figure 2.8 Linked bar chart

### 2.6.3 Progress Bar Chart

Monitoring of activities can take place on the programmes by drawing a 'progress' line parallel to the programme line. The length of the line corresponds to the amount of time spent on that particular activity, as shown in figure 2.9. The problem with this type of monitoring is that the progress line records the amount of time spent on that particular activity not the amount of work completed and so a 'progress column' is sometimes used to record the percentage of completion, as shown in figure 2.10. It can be seen in figure 2.10 that whilst the progress line indicates that excavation has been in progress for one and a half days, three quarters of the time allowed, only 40% of the work is complete. Thus whilst in figure 2.9 the excavation appears to be on time it is in fact about half a day late.

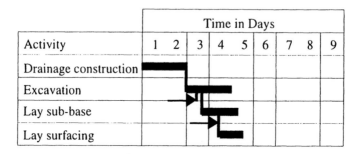

| | Time in Days | | | | | | | | |
|---|---|---|---|---|---|---|---|---|---|
| Activity | 1 | 2 | 3 | 4 | 5 | 6 | 7 | 8 | 9 |
| Drainage construction | | | | | | | | | |
| Excavation | | | | | | | | | |
| Lay sub-base | | | | | | | | | |
| Lay surfacing | | | | | | | | | |

Figure 2.9 Linked bar chart showing progress

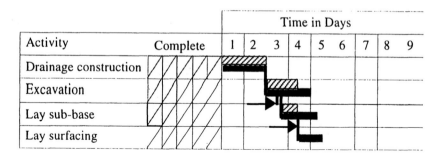

Figure 2.10 Bar chart showing progress as a percentage

Bar charts are used both for organising and monitoring construction activity they can also be used for planning and ordering the resources needed for the job. The blank bar chart layout shown on figure 2.11 has a resource chart at the bottom. This is used to note the number of men, amount of materials and number of machines needed to carry out the activities shown on the Bar Chart. The most efficient way to use resources is under a steady-state condition and not a stop/go situation. For example, it may be more efficient to use one drainage gang at a constant rate for four weeks than it is to use two drainage gangs for one week and two drainage gangs on another week. This type of misallocation of resources can often occur to suit operational matters but will, on its own, be uneconomic due to doubled costs or paying the men to do another activity less efficiently in the intervening week. The resource section of this chart is used to optimise resources and so fine-tune the order of operations to maximise profit. Having optimised Plant and Labour the chart can then be used to indicate when to order materials.

### 2.6.4 Linear Programmes

These programmes are best used for road or rail construction where progress can be almost directly linked to length of the works constructed. An example is shown in figure 2.12 in which the horizontal axis represents time and the vertical axis chainage. These can be reversed as shown in figure 2.13. This example is based on the construction of 800 metres of 12 metre wide airfield runway. The runway is constructed of a flexible road construction, surfacing on sub-base with French drains on each side. Manholes and the primary drainage system have been omitted for clarity. If a structure such as a bridge or culvert is to be constructed at say chainage 700, then this is called a 'static activity' and its programme details are set out in a separate bar chart. In this example the resources available, shown in table 2.1, dictate the rate at which work can be carried out; the programme for the work is shown in figures 2.12 and 2.13.

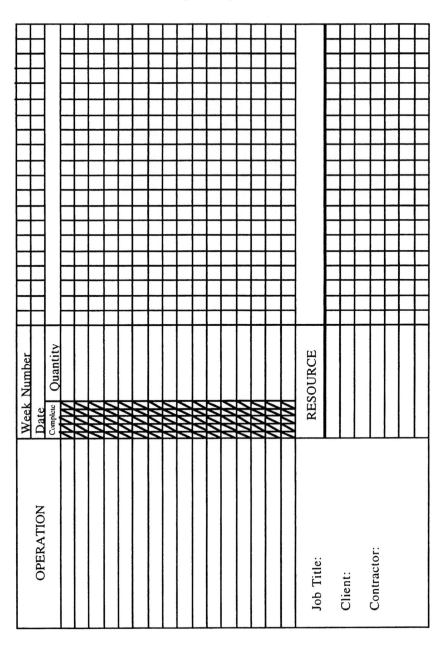

Figure 2.11  Blank bar chart

Table 2.1

| Work item | Resorces | Time |
|---|---|---|
| French Drain | 2 machines<br>(180 and 360 deg Excavator)<br>1 dump truck and 3 men | 80 lin metres/day<br>400m/week (= 200 m of<br>progress along chainage) |
| Excavation | 1 machine, 2 Lorries and<br>1 man | $200 \text{ m}^3/\text{day} = \dfrac{200}{12 \times 0.45}$<br>=37 lin m/day<br>=185 lin m/week |
| Sub-base | 1 machine (+ Roller),1 man | 200 m³/day70 lin m/day |
| Road Base | Sub contract | 100 lin m/day |
| Base Course | Sub contract | 150 lin m/day |
| Wearing Course | Sub contract | 100 lin m/day |

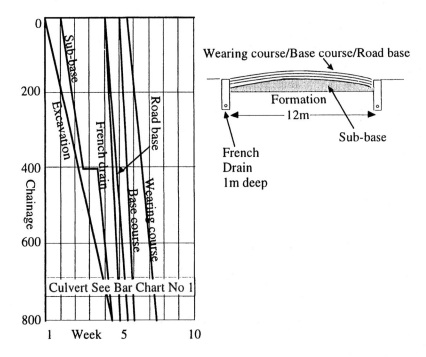

Figure 2.12 Linear chart (with time on horizontal axis) and road cross-section

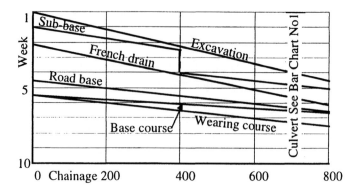

Figure 2.13 Linear programming chart

### 2.6.5  Line of Balance Programme

This type of programme is best used in the manufacturing industry for repetitive tasks, but it can be used in the construction industry for tasks such as house construction or retaining wall construction. For example, consider a 60 metre long flood wall constructed of 10 × 6 metre bays and for simplicity we shall look at the base only; work items are as listed below.

| Work item | Time |
|---|---|
| excavate and blind | = half day |
| steel fix base | = one day |
| concrete | = half day |

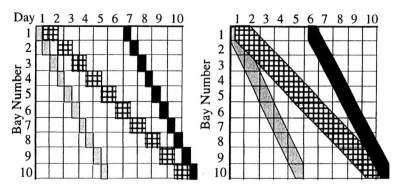

Figure 2.14 Origins of the balance chart

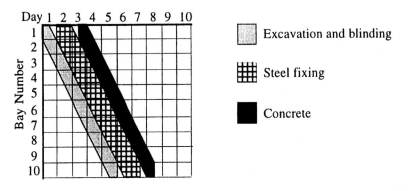

Figure 2.15 Line of balance chart

From figure 2.14 we can see that if production of the second activity, steel fix base, was doubled, time could be saved on the overall completion. So, with the use of two steel fixing gangs (leap-frogging bays) three days are saved on overall completion time. If lines cross or overlap it may mean that two activities are going on in the same place at the same time and this can lead to delays and safety risks. It is best to avoid this situation and so it is important to have a half day buffer between activities, as shown in figure 2.15.

### 2.6.6 Network Analysis

This is a graphical planning technique which shows the project as a network of activities linked together in logical sequence as shown in figure 2.16.

Figure 2.16

The easiest way to explain this system is to look at an example. Consider the construction of the large diameter drain as shown in figure 2.17. The new drain is to connect an existing drainage network into a near by river outfall 45 metres away.

*Sequencing*

To draw a network for this scheme, it is necessary to list each item of work, put them in logical order and then construct the sequence diagram as shown in figure 2.18. This is judged largely from common sense but gets easier with experience and practice. First strip all the top soil over the area of construction, then install the drain pipes. It is good practice to start at the outfall (O/F), the

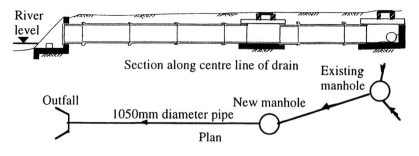

River
level

Section along centre line of drain          Existing
manhole

Outfall                                                New manhole
        1050mm diameter pipe
                                    Plan

Figure 2.17

lowest invert level and work uphill. This is so that any ground water encountered will run away from the area in which work is taking place. To construct the outfall we must put in place a sheet-piled wall to prevent the river from flooding the excavation; any excess water can be pumped out from behind the wall. Sheet piles can be placed whilst the topsoil strip is in process. The pipes can be laid and left open, i.e. not backfilled, until the end of the job. Breaking into the existing manhole (M/H) at the top end can take place after laying all the pipes. The construction of the outfall, the manhole, and breaking into the existing manhole can take place simultaneously because these activities take place in different locations. For both the outfall and the new manhole the following sequence of activities will take place. Excavate bases, place steel and concrete (O/F Base) and (M/H Base), fix steel and shutters for the walls, pour concrete and strike shuttering (O/F Walls) and (M/H Walls). Backfilling of the pipes can then commence whilst the covering slab to the manhole and the tidal flap for the outfall are fixed in place. Backfilling to the outfall must wait until the walls have cured for 3 days and can carry the load. All backfill can now be completed whilst the brick surround and manhole cover are placed. Replace topsoil, harrow and seed with grass. This sequence is shown in figure 2.18.

It is obvious that we cannot strip topsoil and place steel sheet piles at the same time in the same location, so we must stipulate that the topsoil strip in the area of the outfall must be completed first and this is done with a 'forward link'. In this case the forward link states that half a day of topsoil strip must take place before the piling can begin, thus avoiding a clash of activities. In the case of a backward link we specify that the Brick Cover of the Manhole and the O/F Walls must be completed with half a day of backfill activity still to be completed. This will ensure that the backfill activity is started as soon as possible and so keep overall construction time to a minimum.

*Timing*

Table 2.2 shows a list of the estimated resources and time required to carry out each activity. There is inevitably some degree of engineering judgement involved in assessing these times, but it is easier to estimate the time taken for each individual activity than it is to estimate the time taken for the whole job.

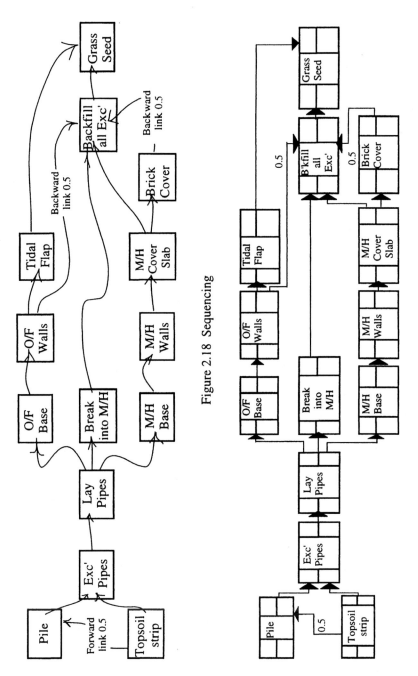

Figure 2.18 Sequencing

Figure 2.19 Network

Table 2.2

| Work Item | Days |
| --- | --- |
| strip all top soil | half day |
| sheet-piled wall | half day |
| pipe excavation | one day |
| pipe laying | three days |
| break into existing manhole | half day |
| outfall base | one day |
| outfall wall and wing walls | five days |
| manhole base | half day |
| manhole concrete walls | one day |
| manhole cover slab | one and a half days |
| brick surround and manhole cover | half a day |
| tidal flap | half a day |
| backfill | two days |
| spread topsoil, harrow and seed | one day |

This is based on experience of a similar activity from a previous job and using judgement to adjust the time in accordance with individual circumstances. Thus, from a series of small decisions we can build up a complete picture of how the job will run. A network of activities can then be drawn up, as shown in figure 2.19.

*Time Analysis*

We have looked at the sequence and then the length of time required for each activity: we must now look at the overall effect of these decisions on the completion time of the job. Each of the activities is drawn as shown in figure 2.20.

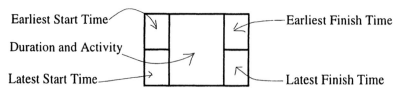

Figure 2.20

Time is usually denoted in days, but weeks or months can be used depending on the detail required. We begin by looking at the earliest start time of each activity starting with 0 at the first activity. This is known as the 'forward pass'. Excavation starts at day 0, the earliest completion time is day 1 and so the earliest start date for the 'Excavation of Pipes' is 0 plus 1, i.e. day 1 and so on. The earliest finish dates are simply the earliest start date plus the length of the activity. For example, earliest finish date for 'Lay Pipes' is 2 + 3 = 5 and so on. The forward pass is shown in figure 2.21.

In figure 2.21 we can see that because of our forward link the piling cannot start until half a day has passed. Each activity then follows sequentially until the activities of 'M/H Base', 'O/F Base' and 'Breaking into M/H' commence on day 5 and carry on simultaneously. The backward links with the backfilling operation mean that the 'O/F Walls' must be complete with half a day backfilling still to go. For this to be the case the Backfilling operation must have started one and a half days before the 'O/F Walls' were completed, i.e. day nine and a half. The Brick Cover activity is shorter than the 'O/F Walls' activity which would indicate that the Backfill activity could commence one and a half days before the Brick Cover activity's earliest finish time, i.e. day seven. This cannot be allowed to occur, however, since it would mean that the backfilling operation would have to wait two and a half days for the O/F Walls to catch up, so we stick with the earliest completion of the 'O/F Walls' by day nine and a half, see figure 2.21. Pathways from 'Break into M/H' and 'M/H cover slab' indicate an earliest start time for backfilling of day five and a half and eight respectively, but these are also disregarded for the same reason.

We then need to consider the latest finish time. This is usually dictated to us by the Client, but for the purpose of this example we will assume that the latest finish time is the same as the earliest finish time, twelve and a half days. We then work back along the network, right to left, filling in the latest start and finish times. Twelve and a half minus one is eleven and a half which is the latest start time for the 'Grass Seed' operation; it is also the latest finish time for the Backfill and so on. This is known as the backward pass and is shown in figure 2.22. Continuing the backward pass along the three pathways brings us to the situation where the Latest Starts are day five for O/F Base, day nine for Break into M/H and day six and a half for M/H Base. The latest allowable finish time for each of the preceding activities must therefore be the earliest of these times.

From the completed network we can gain a number of important pieces of information. First we can tell which are the most important activities, those activities which if they do not start and finish on time will delay the completion of the whole project. We can pick out this sequence of activities as the boxes with the same earliest and latest start time and the same earliest finish and latest finish time. These activities trace out the 'critical path' shown in figure 2.23.

Second we can see that the activities that are not on the critical path have a flexibility as to when they can start and finish; this is known as a 'floating' activity. For example breaking into the existing manhole will only take half a day and can commence as early as day five and not be complete until day nine and a half. This will allow the Site Manager to fine-tune the timing of each operation and use his resources in the most efficient way. It may also allow some flexibility when faced with unexpected problems.

Once we have the network set up we can then start to consider refinements to the working sequence to either shorten the overall contract time or optimise the resources to be used. For example there is potential to shorten the overall completion time of the job by looking at sequencing in more detail. Suppose we allowed the construction of the outfall to commence as soon as the first drain run

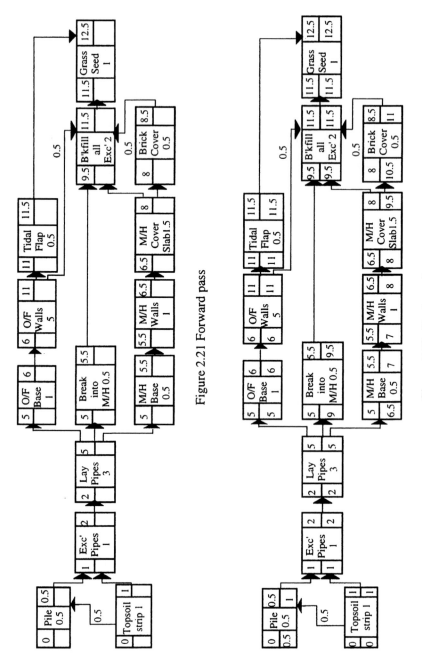

Figure 2.21 Forward pass

Figure 2.22 Backward pass

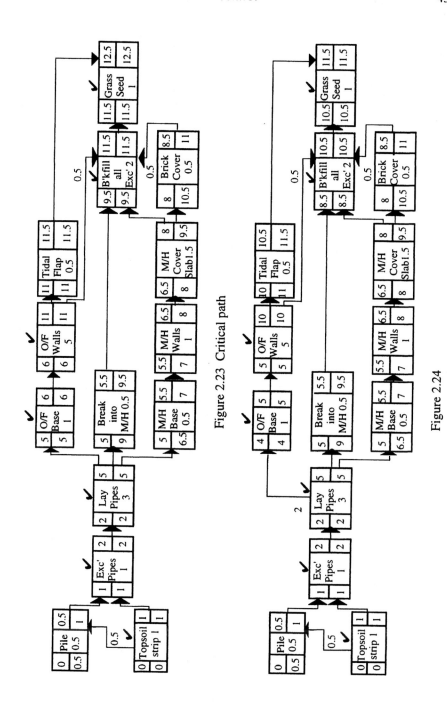

Figure 2.23 Critical path

Figure 2.24

(from O/F to new M/H) is complete. We are introducing a forward link between 'Lay Pipes' and 'O/F Base' of two days. This allows 'O/F Base' construction an earliest start time of day four. Applying this to our network indicates a saving of one day on the overall completion time as shown in figure 2.24.

The advantage of Network Analysis is that it allows precedence, timing and resources to be considered separately and logically. The system allows quite complex construction projects to be analysed based on small sequence decisions and all computer programs are based on this method. Refer to Neale and Neale[4] for further information on this method.

## 2.7 REFERENCES

1.  British Standards Institution. *BS EN ISO 9000 to 9004, 1994, Quality Systems* British Standards Institution.
2.  A. Griffiths. *Quality Assurance in Building,* 1990, Macmillian Press ISBN 0-333-52724-0.
3.  Institution of Civil Engineers, Association of Consulting Engineers and Federation of Civil Engineering Contractors. *ICE Conditions of Contract, 6th Edition, 1991, Conditions of Contract and Forms of Tender, Agreement and Bond for use in connection with works of Civil Engineering Construction.* Thomas Telford. ISBN 0-7277-1617-4
4.  R. H. Neale and D. E. Neal. *Engineering Management - Construction planning,* 1989, Thomas Telford, ISBN 0-7277-1322-1

## 2.8 EXAMPLES

*Example 1*

The data shown in table 2.3 refers to the construction of a six metre bay of two metre high reinforced concrete retaining wall during the summer period. Draw up this data onto a 'Linked Bar Chart Programme.' At least 12 hours must be allowed for the concrete to set. Work in units of days or half days on your programme.

*Example 2*

Consider the construction of a short piece of road as shown in figure 2.25. Access for construction traffic can only be gained from the left hand end. Details of the construction activities and time are as shown in table 2.4. Draw a network of the activities and identify the earliest completion time and the critical path.

Table 2.3

| Work Item | Hours |
|---|---|
| Top soil strip | 1 hr |
| Excavation | 3 hrs |
| Trim | 1 hr |
| Blind | 2 hrs |
| Steel fix base steel | 4 hrs |
| Base shutters | 10 hrs |
| Concrete base | 5 hrs |
| Strip base shutters | 3 hrs |
| Steel fix wall | 2 hrs |
| Make wall shutters | 24 hrs |
| Erect wall shutters | 10 hrs |
| Concrete wall | 5 hrs |
| Strip shutters | 5 hrs |
| Make good shutters | 5 hrs |
| Back fill and trim | 2 hrs |
| Grass seed. | half hr |

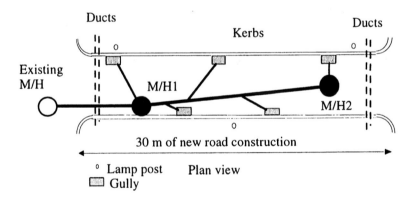

° Lamp post    Plan view
Gully

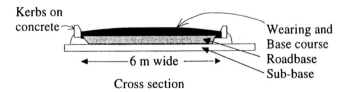

Figure 2.25

Table 2.4

| Work Item | Days |
| --- | --- |
| Top soil strip | Half a day |
| Excavation and trim | One and a Half days |
| Lay sub-base | Three days |
| Place kerbs | Two days |
| Lay roadbase | Three days |
| Surfacing | Three days |
| Manhole Number 1 | Three days |
| Manhole Number 2 | Three days |
| Drainage M/H1 - M/H2 | One and a half days |
| Drainage M/H1 - Existing M/H | One day |
| Install ducts (above drain) | Half a day |
| Install street lighting | One day |
| Topsoil to verge | Two days |
| Seed and make good | One day |

# 3 Safety

The construction industry in Great Britain has one of the worst accident records in the European Community. Although the figures have been falling recently, in 1993/94 65 people were killed and 11,118 people injured. Of the 65 fatalities the causes break down as shown in figure 3.1

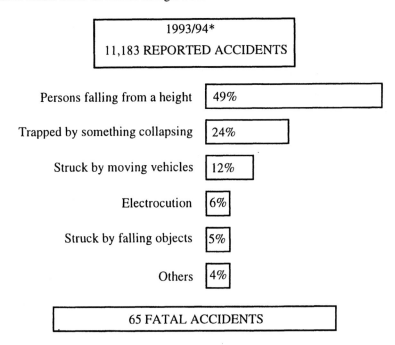

Figure 3.1 Fatalities in the Construction Industry in 1993/4  (*Provisional)
*Information provided by Health and Safety Executive, Statistical Services Unit.*

Of course, this figure varies from year to year depending upon the level of construction activity and strength of the economy, but it is still too high and must be reduced. At the moment any accident which causes a man to lose more than 3 days work must be reported to the Health and Safety Executive or Local Authority. Failure to do so carries a fine of £2000. Such penalties serve to show just how important safety is regarded by government, the law and industry. This chapter begins by looking at health and safety from a historical context and considers some of the major case law judgements which led up to today's emphasis on safety.

## 3.1 STATUTE AND COMMON LAW IN SAFETY MATTERS

Statute Law is imposed as a result of Acts of Parliament. Common Law is decided by judgements made by judges presiding over a case brought before them. Statute Law takes precedence over Common Law and it is true to say that the significance of Common Law in Health and Safety matters has reduced due to the increase in legislation.

Safety was first formally recognised by Statute Law in 1961 by the Factories Act which was brought in to try and improve poor working conditions. This Act, however, did not cover outside work and so a number of judgements were made under Common Law which began to define the responsibilities of persons on site with regard to safety. One such ruling was from Clay v Crump (1964). This ruling clearly stated that all persons on site have responsibility to take care to prevent foreseeable damage.

> In Clay v Crump (1964) the owner of a site retained an Architect to supervise its redevelopment. The redevelopment provided for the removal of a perimeter wall but the Client requested that it remain until the end of the work for security reasons. The Architect consulted the managing director of the Demolition Sub-contractor already on site who considered it safe and they both agreed to allow the wall to be retained.
>
> The Architect visited the site on a number of occasions but did not inspect the wall. After the Demolition Sub-contractors had left site the Main Contractors erected their huts at the base of the wall. During tea one morning the wall collapsed killing two workers and seriously injuring one other. The injured man sued the Main Contractor, the Architect and the Demolition Sub-contractor and all three were held liable.
>
> Damages were awarded against the three at 42% to the Architect, 38% to the Demolition Sub-contractor and 20% to the Main Contractor.

This case alerted the construction industry to its responsibilities that all operatives have a duty of care and was backed up by further rulings, such as Haley v LEB and Baker v Hopkins.

> In Haley v London Electricity Board(LEB) (1965) a blind man (Mr Haley) was walking along the pavement when he fell into a hole dug by the LEB. He was injured and sued LEB. The defence was that a long-handled sledge hammer had been propped up, at an angle, against some railings which was considered adequate warning for sighted people.
>
> The LEB were found liable because it was foreseeable that a blind man might use the pavement.

In Baker v Hopkins (1959) the firm Hopkins was contracted to clean a well 14 metres deep and built a platform half way down on which to operate a petrol driven pump. The managing director of the contracting firm realised the danger that existed from fumes and told his men to leave the pump overnight and not to go down the well the next day until he had arrived on site. The workers, however did go down the well and were overcome with fumes. A Dr Baker was summoned and was urged not to go down the well but he insisted that he must try and help the men down the well. The doctor was overcome with fumes and died on the way to hospital. The Contractor's employees were already dead. The doctor's widow, Mrs Baker, sued the Contractor and they were found to be liable, it being foreseeable that the doctor might attempt a rescue.

## 3.2  HEALTH AND SAFETY AT WORK ACT 1974

In 1974 the Health and Safety at Work Act became law and is still in force to this day. This Act established the Health and Safety Commission and gave it the power to propose health and safety regulations and approve codes of practice. It also set up the Heath and Safety Executive with the responsibility for enforcing Heath and Safety law. This was the first time that construction sites were covered by safety regulations. The intention of the 1974 Act was to place emphasis on individual responsibility for safety. The legislation is aimed at all individuals, employers and employees and not organisations.

### 3.2.1  The Employers Duties under the 1974 Act

The Act requires the following from the Employer:

1. Report all accidents.
2. Display prescribed notices (see figure 3.2).
3. Keep prescribed registers up to date. Registers are kept of checks and inspections of all plant and lifting equipment working on site.
4. Comply with all Laws and Regulations.
5. Provide accident insurance.
6. Provide safety training.
7. Prepare a written safety policy statement (see figure 3.3).

Part of the requirements of the 1974 Act was the necessity to employ a Safety Officer for any company which employed more than 20 operatives. His duties are to assist the company to conform to safety standards and give advice. He is also in charge of safety provision. The Safety Officer's role is advisory and impartial but he may liaise directly with the Safety Inspector (sometimes known as the Factory Inspector).

PRESCRIBED NOTICES

The notices below should be displayed on site, at the employer's office, yard or shop or any place where employees attend and may be able to easily read.

1. Prescribed abstract from the Factories Act (Form 3). On the notice should be shown the name and address of the following:
   District and Superintendent Inspector of Factories
   Appointed Factory Doctor
   Safety Supervisors
2. Copies of the Construction Regulations.
3. The Woodworking Machinery Regulations (if applicable).
4. The Electricity (Factories Act) Special Regulations.
5. The Lead Paint Regulations(on any site with more than 12 persons employed in painting.
6. Electric Shock Placard (if applicable).

Figure 3.2

1. GENERAL POLICY STATEMENT
Stating the policy to safeguard the health, safety and welfare of all employees while at work together with, where practical, people not in the firm's employment, so that they are not exposed to risk.

2. OPERATION OF THE SAFETY POLICY
This section would indicate the responsibilities of all those belonging to the company from management down, and ensure that safety training and instructions are continually provided.

3. ORGANISATION AND ARRANGEMENTS
Specifies responsibility for all matters regarding health and safety, including provision for specific types of work, and the procedure for gathering and communicating new legislation, codes of practice, etc.

4. INFORMATION ARRANGEMENTS
This section is designed to monitor developments in safety, and indicate the type of information which should be collected, e.g. accident reports, developing training material, etc. and together with inspection of places of work, plant transport and equipment, making sure that at all times Statutory Regulations are being observed.

Figure 3.3

### 3.2.2 Employees' Duties under the Act

Employees (sometimes called Operatives) are required to co-operate with the Provisions of the 1974 Act. Each operative must use common sense and:

- Wear a safety helmet at all times (helmets must be manufactured to approved British Standard and be no more than two years old)
- Wear safety boots (This is law since January 1992)
- Use safety equipment provided (such as breathing equipment in manholes)
- Ensure he does not endanger other people in his own work (including the public)
- Co-operate with training.

### 3.2.3 The Safety Inspector

The Health and Safety at Work Act (1974) is enforced by the Health & Safety Commission and the Health & Safety Executive. The Commission appoints Safety Inspectors whilst the Executive pursues legal action in the law courts against companies or individuals who contravene the Act. Safety Inspectors are individuals who have the right to:

1. Examine any site at any reasonable time.
2. Inspect any register.
3. Seek information from operatives relevant to an investigation.
4. Be accompanied by a Police Officer.
5. Remove samples for evidence.
6. Take photographs and measurements.
7. Compel operatives to make statements.

Of particular interest is item 7 since this implies that the Safety Inspectors have more powers than the local constabulary. The Inspector will issue an Improvement Notice if life is not in danger and it is reasonable that improvements can be made. The Inspector can demand that the necessary improvements are made within a specified time. If in the opinion of the Inspector the activities risk serious personal injury a Prohibition Notice will be issued. A Prohibition Notice means that work must stop until the Inspector considers the danger is corrected. He may return within two or three days to lift the Prohibition Notice.

### 3.2.4 The Act in Operation

The Health and Safety at Work Act has worked very well in raising awareness of safety, but there is a lot of work yet to do to reduce the poor record of the construction industry. An example of this is a recent case brought by the HSE against Kenchington Little (1991). During the refurbishment of an existing factory in Nottingham city centre cracks appeared in the facade supported on temporary works designed by the Contractor. These were brought to the attention of the Structural Engineer when on site who did not consider them significant. During high winds on the 2nd February 1990 the facade collapsed. Kenchington Little were found liable and fined £20,000 with costs of £75,000. Consultants argue that because they are on site less frequently than the Contractor, they are less liable for safety matters, but this ruling clearly shows that even if that is the case consultants may still be liable for safety.

## 3.3  CONTROL OF SUBSTANCES HAZARDOUS TO HEALTH 1989

In 1989 the Control of Substances Hazardous to Health (COSHH) regulations came into force. These regulations impose new requirements for the storage and use of chemicals. This act makes it law that the makers' recommendations for the handling and use of chemicals must be followed and requires that a safety policy is drawn up by each Contractor with such matters addressed. Contractors are also required to carry out a Hazardous Chemicals audit and to seek specialist advice on any matters they are unsure of. For further information see Report 125 by CIRIA[1].

## 3.4  EUROPEAN LAW

In January 1973 the European Communities Act became law in this country which effectively made the United Kingdom part of the European Union and subject to European Law and Directives. The single European Act introduced the requirement into the Treaty of Rome that member states 'pay particular attention to encouraging improvements, especially in the working environment, as regards the health and safety of workers'. A number of directives from the European Union have been introduced into UK law via 1974 Health and Safety at Work Act and these include the following which directly affect the construction industry.

- Manual Handling Operations Regulations 1992.
- Personal Protective Equipment at Work Regulations 1992.
- The new Construction Design and Management Regulations (CDM) 1994.

### 3.4.1  Manual Handling Operations Regulations 1992

In 1992 the Manual Handling Operations Regulations came into force with effect from January 1993. These cover all areas of employment, and require that employers identify all hazards and risks associated with normal working activities and take action to reduce those risks.

### 3.4.2  Personal Protective Equipment at Work Regulations 1992

The Personal Protective Equipment at Work Regulations 1992 came into force in June 1995. These regulations impose a duty on the employer to assess the risk and make provision for working in all environments, for example provide wet weather gear for inclement weather working and head and ear protection as necessary.

### 3.4.3 The New Construction Design and Management Regulations 1994

In 1994 The new Construction Design and Management Regulations (CDM) were made law and came into effect on 31 March 1995. The main thrust of this legislation is the extension of responsibilities for Health and Safety matters to Clients, Engineers and Designers. Designers are required to take into account Health and Safety matters in the construction of their work implied in the design and in the materials they specify. Clients are required to appoint a 'Planning Supervisor' who will ensure that the Designers have taken due regard of Health and Safety and that the Contractors are aware of the risks involved in the project.

*The Client*

The Client must appoint a Planning Supervisor who is competent to carry out the duties involved. He must also make sufficient resources available to the Planning Supervisor and Contractor to enable them to carry out their duties.

*The Planning Supervisor*

The Planning Supervisor will oversee the whole construction process from conception to completion. He will ensure that a health and safety plan is prepared, monitor the design with regard to health and safety, give advice to the Client, Contractor and Designer and ensure that a Safety File is kept on the whole project on specific matters of health and safety. The responsibilities of the Planning Supervisor may be transferred to different persons employed by the Designer, Contractor and Client's organisation appropriate to the phase of existence of the facility but the post must be filled until demolition.

*Designer*

The Designer must make the Client aware of his duties under the Regulations. The Designer must ensure that his designs avoid unnecessary risk to health and safety in both construction and maintenance of the project. The Designer will be required to act on advice offered by the Planning Supervisor that may reduce the risks to Health and Safety.

*Principal Contractor*

In addition to his duties under existing Health and Safety Regulations the Contractor is required to develop a health and safety plan. The plan is intended to co-ordinate the efforts of the parties to the contract to improve matters relating to health and safety. Such a plan might predetermine the dates of Health and Safety Meetings and define the duties of those who attend.

For a fuller explanation of Construction Design and Management Regulations see *The CDM Regulations Explained* by Raymond Joyce[3].

## 3.5 NEW ENGINEERING CONTRACT

The New Engineering Contract[2] requires the provision of a Safety Statement, the submission of safety policies for inspection and a statement defining which party to the contract is responsible for maintaining areas on site which may be occupied by several Contractors at the same time.

## 3.6 CASE STUDIES

### 3.6.1 Case Study 1

The Health and Safety Executive v British Rail and the Health and Safety Executive v Tilbury Douglas.

On 13th June 1992 two spans of a Victorian three arch brick bridge at St Johns' Station, London collapsed during demolition killing two men. The Main Contractor was Tilbury Douglas and the Engineer British Rail. The method statement for demolition, approved by British Rail was to evenly excavate the top of the bridge with two 360° excavators until the centre arch could be removed. No mention was made of bracing of the outer arches during this operation in which case instability is inevitable. In the event excavation was not carried out evenly and the north machine substantially completed its excavation whilst the south machine had been delayed. Inevitably, collapse occurred due to instability, killing two men working under the bridge. The Health and Safety Executive took action in the courts against BR and the main Contractor Tilbury Douglas. British Rail pleaded guilty to 'Failing to ensure the safety of people not in their employment', contrary to section 3 of the Health and Safety at Work Act. Tilbury Douglas pleaded guilty to the same section and to section 2 of failing to ensure the safety of its own employees. Both were fined £25,000.

### 3.6.1 Case Study 2

Here is a hypothetical situation for discussion:

You are the project manager of a construction site in London. The construction site has a number of deep excavations which remained open during the two-week Christmas break. The site is fully enclosed by two metre high plywood boarding with two observation ports cut into the fence. Access to the site is via double gates of plywood (two metre high) which were padlocked during the shutdown. The excavations are not fenced off.

a) During the shutdown, some children gained access to the site via an observation port, one of whom fell down a deep excavation and was injured. The parents of the child sued the Contractor and Engineer. Should the complaint be upheld or not, give your reasons?
b) It is revealed that the Engineer asked the Contractor to board up the observation ports before Christmas. This was not done.
If this changes your judgement, briefly indicate your reasoning.

## 3.7 REFERENCES

1. Construction Industry Research and Information Association (CIRIA), *Report 125 'A Guide to the Control of Substances Hazardous to Health in Design and Construction'*, 1989, Thomas Telford ISBN 0-7277-1979-3
2. Institution of Civil Engineers, *New Engineering Contract*, 1990, Thomas Telford ISBN 0-7277-1616-6
3. Raymond Joyce, *The CDM Regulations Explained*, 1995, Thomas Telford ISBN 0-7277-2034-1

# 4 Ground Water Control

Constructing works below ground level is a major part of the work of a Civil Engineer and the ingress of ground water can cause serious problems. Water-logged conditions can slow work considerably and cause safety problems, such as erosion or collapse of the sides of an excavation, or even flooding, and so ground water must be controlled to carry out work effectively. It is not possible to consider ground water control without also looking at ground support methods since many systems have a dual purpose. Such systems will be explained within the text. In general there are two types of water to control.

- Surface water which is the result of rain collecting in hollows and depressions both on the ground surface and immediately below. Any surface water collecting in a depression in the layers of soil below ground is sometimes called a 'perched water table'.
- Ground water which is the water naturally contained within the soil up to the water table level.

This chapter looks at a comprehensive range of ground water control techniques in both the temporary and permanent conditions. The explanations are aimed at drawing together aspects of both design and construction. The chapter concludes with an overview of methods and case studies.

## 4.1 INTRODUCTION

The effect of the water on the soil depends upon its 'particle size' and 'permeability'. For simplicity we shall consider the two extremes of soil type: clay, which consists of very fine particles and is of low permeability, and gravel, which has a large particle size and is highly permeable.

Clay is impermeable (water cannot penetrate clay very easily) and so water is not too much of a problem in temporary excavations, provided that the sides of the excavation can be made stable by some form of propping or by sloping at a safe angle (see Chapter 5 on Earthworks). Any collection of ground water or surface water can be pumped away. If water is allowed to stand on the surface of clay, however, the movement of construction vehicles will cause ruts and considerably weaken the clay, making it in most cases unfit to use as an engineering material. Long term stability of clay is more complicated because water may permeate into or out of the clay over a period of time, say 1 or 2 years. The volume and strength of clay depend on the water content. The more water in a clay, the greater the volume and the lower the strength; with less water content the clay contracts and the strength increases, and for the engineer this is a fundamental property of the material. We are interested in the strength of clay as an engineering material; for example, to construct an embankment, or cutting, or stand a building on it and we can use drainage to control water content and thus the strength condition.

In the case of gravel, water can penetrate very easily since the gaps between the particles are large enough to allow water to pass. The volume and strength of gravel do not depend on water content and so permanent and temporary loading conditions can be treated in the same way for design. The movement of ground water, however, may cause problems by washing away fines (fine material) and forming voids within the soil. If voids are formed the soil will be greatly weakened and may even collapse. Gravel must be contained, so that it can develop strength. A casual walk along the beach will show how gravel behaves when unconfined, it ruts and 'rolls' very easily. A lorry load of gravel when tipped on the ground will take up a natural pile shape with sloping sides. The angle of the slope to the horizontal is known as the 'natural angle of repose'. Excavations carried out in gravel must, therefore, either be supported against collapse or allowed to slope at some angle slightly less than the natural angle of repose. An excavation which is allowed to slope without support is known as an 'open cast' excavation.

From the above we can see that the control of ground water is very important in terms of maintaining the strength of the material in both the permanent and temporary conditions. Ground water and surface water can be controlled in a number of ways, but the answer to the problem depends on the requirement of either temporary or permanent solutions.

• Temporary solutions:

Sump pumping and grips
Sheet piling/cofferdams
Well point system
Deep bored wells and pumping
Horizontal drains
Electro-osmosis
Ground freezing

• Permanent solutions:

Drainage/sand drains
Sheet piling (permanent and temporary)
Diaphragm walls
Slurry trench cut off
Thin grouted membrane
Contiguous piling
Pressure grouting

Methods of ground water control can also be categorised in terms of dewatering systems or a water exclusion systems. Dewatering means the lowering of the water table by pumping the water out of the ground until its level is below the level of the intended excavation; here water is continually moving within the ground. Water exclusion is the construction of a barrier which will stop the water getting into the proposed excavation.

## 4.2 TEMPORARY SOLUTIONS

### 4.2.1 Sump Pumping

This temporary solution recognises that water will get into the excavation but it is pumped away to prevent flooding and maintain access. Sump pumping is suitable for most soils but can only be used for excavations of up to five metres in depth (see figure 4.1). The depth of the excavation is limited by the suction lift of the pump. The pump must 'suck' the water up the inlet pipe to the full height of the excavation; this requires air-tight joints which are not practical to achieve. The sump must be lower than the general level of the excavation to prevent water from standing on the bottom of the excavation and keep the inlet pipe submerged. The bottom of the excavation should be provided with a small gradient, falling towards the sump to avoid standing water.

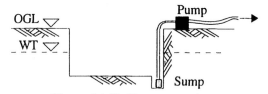

Figure 4.1 Shallow excavation

In a situation when there is a large area to drain or the excavation is expected to remain open for some time, it may be necessary to dig small channels filled with gravel called 'grips' falling towards the sump or a peripheral grip around the edge of the excavation.

The pumps which are used fall into the two broad categories of piston and centrifugal. Piston pumps work because a one-way valve system allows water into the piston chamber on the upward stroke and out of the piston chamber via another valve on the downward stroke. Two such pistons are usually operated together such that one piston is always on the upward stroke and is classified as a double-acting pump. These pumps are of low capacity up to 3600 litres per hour and work at low pressures, but have the advantage of high efficiency regardless of head and speed of operation. See figure 4.2. Centrifugal pumps on the other hand are high volume, high pressure pumps. They have an output of between 3600 and 90000 litres per hour at a pressure of between 140 and 1000 kN/m². They work by a rotating impeller, immersed in water, forcing the water round at high speed against the outer manifold which contains the exit valve.

The stability of the sides of the excavation is a very important factor to be considered when designing the water control measures. If the soil is a fine sand or silt, then the very act of pumping water out from the sump may produce collapse of the sides of the excavation. This is caused by the movement of the water in the soil, induced by the pumping. To prevent this the excavation must be supported. Support is often achieved by using steel sheet piles.

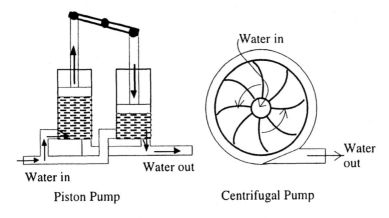

Water in

Water out

Water out

Water in

Piston Pump          Centrifugal Pump

Figure 4.2

## 4.2.2 Steel Sheet Piling

This form of piling is the most commonly used for permanent and temporary work to control ground water and support excavations. As permanent retaining walls it is used for harbours, docks, river banks and sea defence work. Steel sheet piles are also used in the temporary case to support the sides of trenches and in cofferdam construction. The sheet piles interlock to form a steel wall which in addition to supporting the soil can also provide a partial barrier against the ingress of water. Steel piles are usually hammered into the ground and once in the ground are held stable by either cantilever action or are propped across the excavation. (See Chapter 7 on Retaining Walls)

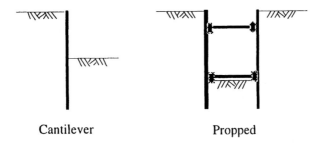

Cantilever          Propped

Figure 4.3

Propping is accomplished either by using trench props at design centres bearing onto a continuous beam called a 'waling', or in the case of a cofferdam by a steel frame suitably designed to resist the loads. The disadvantage of using this type of support is that it restricts access, especially of excavation equipment and so an alternative form of support is derived from ground anchors. (See Chapter 7 on Retaining Walls)

In the UK steel piles are produced by British Steel who produce two types of patented pile, the Frodingham and the Larssen. Cross-sections of each are shown in figure 4.4.

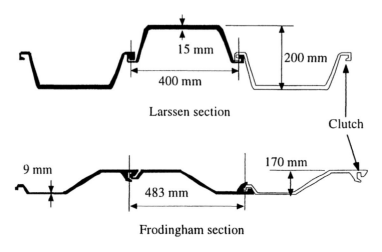

Larssen section

Clutch

Frodingham section

Figure 4.4

These piles are rolled in lengths from 9 to 30 metres. To install the piles they can be hammered in, pushed in or vibrated into position. Whatever method is used the piles must be guided into the ground by a piling frame. The piling frame ensures that the piles go in plumb and that the clutch on either edge is engaged with the clutch of the adjacent pile. The piling frame provides two walings at the top and two near the ground to provide the necessary support at about three to four metre centres. It is essential that the walings are held securely and this can be achieved in many ways, but usually some form of braced frame is used. See figure 4.5. Steel piles must be driven in 'panels', that is, the number of piles that will fit in the piling frame at any one time. Piles must also be driven in pairs and in a specific sequence to ensure that they do not 'lean' on installation and that the clutches remain engaged. The first two piles are pitched and piled half way, taking great care to maintain verticality, until the tops of the piles are level with the top of the piling frame. Pitching is the term used to describe the operation where the piles are lifted into the frame with the end resting on the ground and with the clutches engaged with neighbouring piles. The rest of the piles which form the panel are then pitched and the two end piles are driven half way until the tops are at the top of the frame. Again, great care must be taken to ensure that these piles are plumb. The rest of the piles in the panel can then be driven in pairs to the same level. This sequence is shown in figure 4.5.

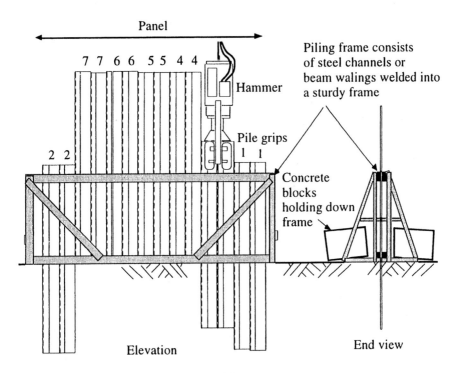

Figure 4.5 Piling sequence

The piling frame is then removed and the same sequence is followed to pile the next panel to the required depth. The first panel is then completed by piling to the required level. Steel sheet piles can be used in many soil types but it is often not economic unless they can be reused. There are three methods used for the insertion of piles. The conventional method is by diesel hammer which drives a heavy weight on to the pile head using the combustion cycle of the fuel. It is noisy and so not popular in built-up areas, but it is fast and effective. The hydraulic system uses eight hydraulic rams, two connected to the top of each of four piles. One pair of rams pushes one pile in at a time using the other three to pull or push against. A set of four piles are thus installed in small steps and it is a very quiet system producing a maximum of 75 dBA. The vibration system simply vibrates the pile at high frequency which 'liquefies' the ground on contact. The pile is then pushed into the ground under its own weight. This system is best used with a silt or sandy soil. It is quick, quieter than a diesel hammer, but noisier than the hydraulic system.

### 4.2.3 Steel Pile Design

Once the piles are installed, they form an effective barrier against ground water or free flowing water if installed as part of a cofferdam and will support the sides of the excavation. Steel sheet piles are not normally water-tight and pumping is usually required. When excavation begins the piles will be subject to the lateral pressure of the water and the soil that they are retaining. The pressure from water is called 'hydrostatic' pressure. The pressure from the soil is called the 'active pressure' and is given by Rankine's equation.

Active pressure $P = K_a \gamma h.$          Hydrostatic pressure $= \gamma h.$

Where:
   $K_a$ = Rankine's Constant for active pressure, usually 0.33 for granular soil.
   $K_p$ = Rankine's Constant for passive pressure, usually 3.0 for granular soil.
   $\gamma$ = Density of the soil or water.
   h  = Depth below the surface.

In a cantilever design the active ground pressures are trying to push the pile over, but they are resisted by the passive ground pressures which are trying to maintain the status quo. This pressure is called the 'passive pressure' and is estimated using Rankine's equation as above. The pressure distribution is as shown in figure 4.6.

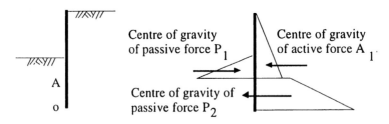

Figure 4.6

For design, we must first check for overturning stability, i.e. check that the system of pressures and forces will not push the pile over. Secondly, we must check that the pile will not fail in bending, i.e. check that the pile will not bend, allowing the retained soil to collapse. The stability check assumes a simplified system of forces as shown in figure 4.7.

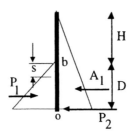

Figure 4.7

simplification is allowed for by increasing the designed length by 20 % at the end of the calculation. Taking moments about 'o' we have the following.

$$P_1 \times \frac{D}{3} = A_1 \left(\frac{H + D}{3}\right) \tag{4.1}$$

where $P_1 = 0.5K_p\gamma D^2$ and $A_1 = 0.5K_a\gamma (H + D)^2$

For the pile to be safe, the left hand side of equation (4.1) has to be greater than the right hand side by a factor of two. If this is the case the system is said to have a 'Factor of Safety' (FOS) of 2. To achieve this we simply guess values for D and try it in Equation 1 until we achieve a FOS of 2; with experience this becomes quicker and thus we can find the length of pile necessary for stability.

We must now check that the pile will not bend over under the pressure of the soil retained. The section properties of the pile determine how much bending the piles can resist and this must be compared to the maximum applied bending moment. Maximum bending moment occurs at the point of zero shear, shown as s below b in figure 4.7.

Equating forces $0.5K_p\gamma s^2 = 0.5K_a\gamma (H + s)^2$ but $K_a = \dfrac{1}{K_p}$

so $(K_p)^2 s^2 = (H + s)^2$ which equals $K_p s = H + s$

therefore

$$s = \frac{H}{K_p - 1} \tag{4.2}$$

The maximum bending moment can then be found by taking moments about the point of zero shear. The Engineer must then go on to check that the deflection at the top of the pile is acceptable to the user but that is beyond the scope of this book. For further information, refer to CIRIA Publication 95[3].

*Example*

A cantilever sheet piled wall is required to support an excavation three metres deep. Soil is granular with a density of 20 kN/m³, $K_a = 0.33$ and $K_p = 3.0$. Find the required pile length. The water pressure is ignored for simplicity but this must be taken into account.

$P_1 = 0.5 \times 3.0 \times 20 \times D^2$ and $A_1 = 0.5 \times 0.33 \times 20 \times (3 + D)^2$

If $D = 3$ say, then $P_1 = 270$ and $A_1 = 119$

Left hand side of equation (4.1) is $270 \times 1 = 270$

Right hand side of equation (4.1) is $119 \times 2 = 238$

$$\therefore \; FOS = \frac{270}{238} = 1.13 \text{ this is less than 2 and so the depth must be increased.}$$

Try     D = 5     then     $P_1 = 750$     and     $A_1 = 211$

Left hand side of equation (4.1) is $750 \times \frac{5}{3} = 1250$

Right hand side of equation (4.1) is $211 \times \frac{8}{3} = 563$

$$\therefore \; FOS = \frac{1250}{563} = 2.22 \qquad \text{this is greater than 2}$$

To allow for the simplified model, we must increase this by 20% so that the pile length required is 5 × 1.2 + 3 = 9 m: but we have not taken account of hydrostatic pressure, the buoyancy effect on the weight of the soil, or the flow of water within the soil under the piles so this cannot be considered as a complete design. Now we need to find the applied moment.

From equation (4.2)         $s = \frac{3}{2.3} = 1.3$ m

Maximum bending moment $M = 0.3 \times 20 \times \dfrac{(H + s)^3}{6} - 3.3 \times 20 \times \dfrac{s^3}{6}$

$$\therefore \; M = 0.3 \times 20 \times \frac{4.3^3}{6} - 3.3 \times 20 \times \frac{1.3^3}{6} = 55.3 \text{ kNm}$$

### 4.2.4 Cofferdams

Cofferdams are used for deep confined excavation in poor ground or in open water; a general arrangement is shown in figure 4.9. The steel piles and internal frames must be designed to resist the forces from the soil and the ground water.

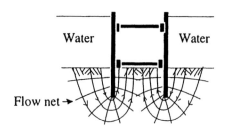

Figure 4.8
The cofferdam is designed for overturning stability and for the 'quick'

The cofferdam is designed for overturning stability and for the 'quick' condition, i.e. quick sand. Ground water will seep under the piles through the ground, at a rate which depends upon the permeability of the soil and the head of water to be retained. The curved path followed by the water is called the 'flow net', see figure 4.8. On the upward stage of the journey the movement of the water causes a greater buoyancy effect on the particles of soil. The buoyancy effect becomes significant when the upward force exerted on the soil particles is equal to or greater than their weight. In these circumstances the soil particles become suspended and the soil behaves as a fluid. This is known as 'boiling' or as 'piping'. Under such conditions the soil cannot support load and will be pushed up the inside of the excavation by its own buoyancy, engulfing anything in its path. Once the water head is equalised the soil will assume a stable state. Such disasters can be avoided by lengthening the piles to extend the length of the flow net or by casting a heavy concrete base at the bottom of the excavation to attempt to equalise ground pressures. For further information on this topic refer to Whitlow[2] and Barnes[1].

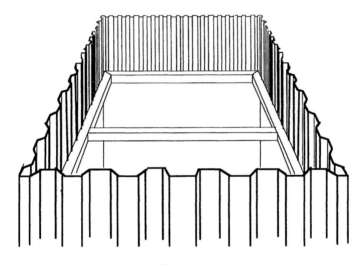

Figure 4.9

### 4.2.5 Well Point System

Steel piles are used to control ground water by forming a barrier across which it will be difficult for water to pass. Well point systems, however, approach the problem from a different perspective. The well point system works on the principle of lowering the surrounding water table, to a point below the intended level of excavation. The process is called dewatering and, if carried out before excavation, the dig is said to be carried out in 'dry' conditions. The ground may of course still require support if confined excavations are required, but open cast excavations can be carried out without further support. Suitable for most granular materials and cheap, the well point system is the most popular system of ground water control used today. The well pipe is first jetted into the ground by forcing water out of the end of the pipe and liquefying the ground. The pump usually used for this work will be a centrifugal pump with an output of about 75000 litres per hour at a pressure of approximately 700 kN/m². This mobilises the soil particles by forming the 'quick condition' and the pipe sinks into the ground under its own weight. Once at the required depth the water pumping is gradually slowed to a stop and the soil is allowed to settle. The slowing process premits the soil to settle in a segregated way, allowing the larger particles to settle close to the well point and finer particles further away, which helps the soil to retain its finer particles and prevents them from being sucked up into the pump. Once the soil is settled the pump is reversed so that the pipe now acts as a well and ground water can be extracted. This process is shown in figures 4.10 and 4.11. Some 15 to 20 well points can be operated with the same pump.

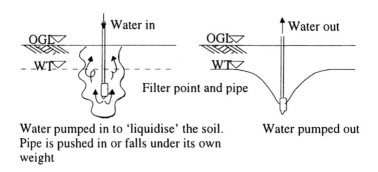

Water pumped in to 'liquidise' the soil. Pipe is pushed in or falls under its own weight

Water pumped out

Figure 4.10                    Figure 4.11

Any voids remaining around the well point riser pipe are backfilled with a granular material. The well point head is a small but important piece of equipment. It consists of a slotted outer tube about 1.2 m long covered with a fine mesh as a barrier against the intrusion of fines. An inner tube about 38 mm in diameter has a jetting valve at the tip to allow water to pass out under pressure for jetting in and for pumping out. The inner tube is connected to the riser pipe and then to the header pipe via a swing pipe. See figure 4.12.

Disposable well point heads can now be obtained giving the obvious advantage of economy in the event of loss or damage to a head. The header pipe is a high pressure flexible hose 100 mm in diameter which lies along the surface receiving well point risers at about 1.2 m intervals. The header pipe is connected directly to a high pressure pump.

This dewatering system is usually installed before excavation is allowed to commence, thus avoiding risk of ground movement as a result of draw down of the phreatic surface when pumping begins. As the ground water is removed the surface of the water table, called the 'phreatic surface' is deformed into a conical depression. The diameter of the conical depression depends upon the permeability of the soil; it will be larger for more permeable soils. The pattern of deployment of the well points must be considered carefully, taking due regard of the type of soil and the depth of excavation. Pumping will have to be continuous to maintain the reduced levels of ground water. For a confined excavation a ring main pattern is adopted around the top of the excavation; for a trench a rolling system is used. A rolling system simply uses three sections of well points, one for withdrawal, one in operation where pipe laying is in progress and one for installation, leapfrogging along the trench line in pace with the rate of construction. If a clay/silt is encountered, well points may be required on both sides of the trench.

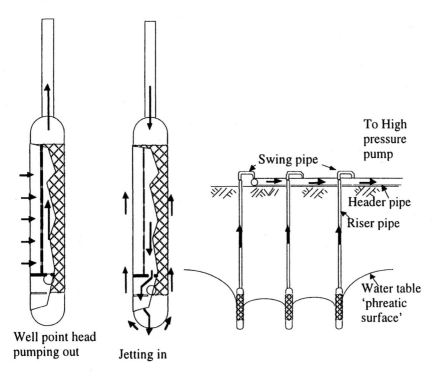

Figure 4.12

The depth of penetration is limited to the lift height of the pump, usually five metres, so if excavations deeper than this are required then a multi-stage system can be used in conjunction with terraced excavations. Care must be taken in this situation to design the stability of the embankment to allow for the effect of moving water within the slip circle induced by the dewatering system. All embankments have a circle of potential failure which must be checked by calculation to ensure safety; this is called the 'slip circle'. Slip circle calculations must take account of forces resulting from the seepage of water being extracted. (See Whitlow[2] for details of slip circles). Deep bored wells could be used at the top of an embankment to reduce the seepage problem. See figure 4.13.

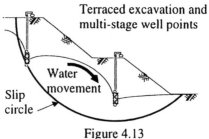

Terraced excavation and
multi-stage well points

Figure 4.13

### 4.2.6 Deep Bored Wells

If the ground is not suitable for the well point system, or if depths in excess of five metres are required, deep bored wells may be employed, as shown in figure 4.14. Here a pump is lowered to the bottom of the deep well to dewater the ground. The hole is bored with conventional drilling or boring machines and a steel lining is left in place of 300 to 600 mm in diameter. Another steel tube is placed inside the well called the inner lining which is perforated and in which a submersible pump will be placed. The space between the two linings is filled with a permeable material. Before the pump is placed in position the well is purged by forcing water down the inner lining and out though the perforations, granular fill and outer lining to wash out unwanted fines and allow the free flow of water into the well.

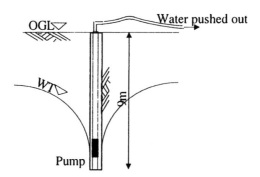

Figure 4.14

When the pump is in operation a conical depression forms in the water table as it does for well point system. The radius of this depression and so the spacing of wells depend on the permeability of the soil and since this system is obviously more expensive and slow to install the maximum use has to be made of each well. To increase the effect of the well on the surrounding water table horizontal drains can be installed radially from the well below water table level. This will have the effect of increasing the zone of influence on the water table and of thus increasing the rate at which water can be removed.

### 4.2.7 Ground Freezing

This method has been used extensively in tunnelling to construct shafts as deep as 200 m and in driving tunnels to control ground water. It can be used to provide a waterproof barrier in poor ground whilst at the same time stabilising the sides of the excavation. The method consists of steel freeze pipes 100 mm in diameter and as long as is required placed in the soil at about one metre centres. A heavy solution of brine (containing calcium chloride) is circulated through the pipes to a refrigeration unit which can be lorry mounted as shown in figure 4.15.

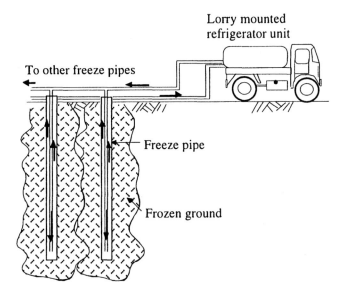

Figure 4.15

The refrigeration unit maintains a temperature of between $-15°$ and $-25°$ C and circulates the brine through the pipes causing the surrounding soil to freeze. Freeze pipes at one metre centres will produce a wall of ice in the ground one metre thick in sand and gravel in about 10 days of operation, but clay would take 17 days. In an emergency liquid nitrogen could be used which would freeze the ground in a few days.

This system of ground water control is competitive with other systems below depths of nine metres and it becomes cheaper as the depth increases. The only limit to the depth of effectiveness is the length of the freeze pipes. Freezing can be used in soil with moisture contents as low as 10 per cent and the rate of freezing can be altered with the pipes at closer centres. This is a water exclusion and support system which relies on a low temperature being maintained in the ground at all times. When first installed care must be taken to ensure that the ground is completely frozen before excavation begins. To this end an observation well can be used to observe the ground water being forced up the hole as the ground freezes. Freezing the ground will cause it to expand as the water crystallises inside the soil structure, which in this context is called 'heave'. A soil is susceptible to heave if it has a predominance of fine particles such as clay or silt. Once affected by heave the soil may become weaker when it has defrosted than it was before freezing. This method must not, therefore, be used in an area which currently supports load or is intended to support load from a structure without considering its effects on the long term strength of the soil.

### 4.2.8  Electro-osmosis

This is an expensive method of dewatering but can be effective to dewater cohesive soils, i.e. clay. All soil particles carry a −ve charge whilst water carries a +ve charge. Within the clay the opposite charges attract, holding the water in place, thus forming a stable structure and slowing down dewatering. This natural balance can be upset by applying a positive electrical charge to the filter of the inlet of the pump and a negative charge to the soil, using a steel probe, figure 4.16.

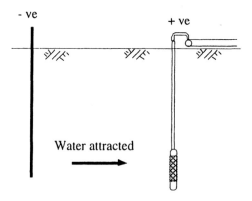

Figure 4.16

The water is then attracted to the inlet of the pump and is pumped away. This method can substantially increase the rate of water extraction but a large amount of electrical power is required and the effect is local.

### 4.2.9  Effects of Lowering the Water Table

Local removal of water from the water table causes water to flow within the soil toward the area of removal. The rate of flow within the soil depends upon the permeability, but in all cases fine grained material may be washed out of the soil structure. In gravel the removal of fines causes cavities which may give rise to local settlement or instability; the severity of the problem depends upon the proportion of fine to coarse material, known as the grading. Any pumping process such as this must be discharged into a settlement tank, so that a check can be made that no fines are being extracted. If they are, the mesh size on the filter of the intake to the pump will need to be changed. Removal of water from the ground may cause changes in the volume of the soil and so it is important to monitor settlement of nearby foundations and ground levels. In clay this effect is particularly marked, since removal of moisture causes it to shrink and addition of moisture causes it to swell or heave. A common cause of settlement in the south-east of England is the removal of moisture from the underlying clay by tree roots.

## 4.3 PERMANENT SOLUTIONS

### 4.3.1 Drainage

This is often the cheapest method of ground water control especially if a pumping station is not required. Here we can use free-flowing, gravity fed French drains, horizontally or sand drains vertically, figure 4.17. French drains are commonly used to protect road formation from ground water. They run along both edges of a road and may even go under the road, keeping the water table well below the formation. This type of drainage is best suited to the control of water over large areas but at shallow depths, up to 1.5 m from the surface.

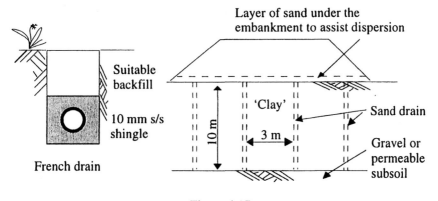

Figure 4.17

Vertical drains are used in clay which we may wish to consolidate, so that it will carry a greater load such as the weight of an embankment. They consist of holes drilled in the ground at close centres and filled with sand or some form of geotextile strip which allows an easy passage for ground water. To consolidate clay it is common practice to overload the ground with the weight of extra soil to squeeze the clay and push the water out of the soil structure. This may take a long time if the clay is thick because the water has a long way to travel. The installation of vertical drains speeds up the evacuation of water by providing an easy passage out of the clay. The journey to a vertical drain through the clay is shorter than to the surface or the underside of the clay layer.

### 4.3.2 Diaphragm Walls

A diaphragm wall is a structure which spans in two directions rather like a drum skin and in this case consists of an in situ concrete wall cast in the ground. It can be used to control the passage of ground water and support excavations. It is particularly useful for the construction of deep basements and is suitable for all types of soils except boulders and rock. Diaphragm walls are usually more expensive than other forms of ground water control and so are only economic as a permanent solution. For this application the walls are cast in the ground in deep trenches up to 30 m deep and between 450 mm and 1 m wide. The excavation is carried out using a clam shovel, see figure 4.20, and continually filled with 'bentonite' to prevent the ingress of ground water and stabilise the excavation from collapse. Concrete can then be placed using a tremie tube to displace the bentonite as the excavation is filled. Reinforcement may also be placed in the concrete if the wall is to perform a structural role, as shown in figure 4.18.

Bentonite (or Fuller's earth) is a thixotropic fluid of specific gravity 1.2, heavier than water and looks like a runny mud slurry; it is brought to site as a powder which is mixed with clean water to produce the fluid. A trench full of bentonite is able to exert enough hydrostatic pressure to prevent the ingress of ground water and prevent the collapse of the sides of the excavation. Bentonite is allowed to flood the guide trench before excavation begins and will naturally flow into any holes made by the clam grab during the operation. The bentonite can also keep the excavation free of dislodged soil by being continuously circulated through a filter system on the surface.

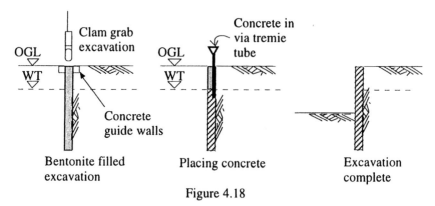

Figure 4.18

A tremie tube is a pipe about 200 mm in diameter which is long enough to reach the bottom of the excavation and has a hopper at the top. Liquid concrete is placed in the hopper and then moves down the pipe under its own weight. (some assistance may be given to this action by vibrating the tube). The concrete comes out at the bottom of the pipe where it accumulates, filling the excavation and displacing the bentonite. The purpose of the tremie tube is to prevent the concrete becoming contaminated with bentonite and to control the rate of fall into the excavation and thus prevent segregation of the concrete mix. Segregation occurs when the concrete is allowed to free-fall distances greater than two metres and on impact the larger stones separate from the cement and fines. Both segregation and contamination with bentonite would weaken the concrete. As the concrete level rises in the excavation, the end of the tremie tube is held just below the surface of the concrete again to prevent contamination. The excavation is filled until the contaminated concrete at the top of the pour over-spills and is removed by pumping it into the next excavation.

A diaphragm wall is constructed in alternate panels usually about five metres long, figure 4.19. This is to allow the concrete to cure and any thermal movement to take place before the adjoining panel is cast, thus keeping cracking to a minimum. Two excavation machines will usually work together and may be guided by shallow concrete ground beams cast at the surface either side of the desired position of the wall.

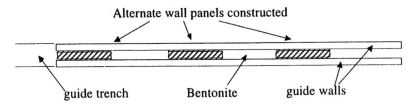

Alternate wall panels constructed

guide trench          Bentonite          guide walls

Figure 4.19 Plan View

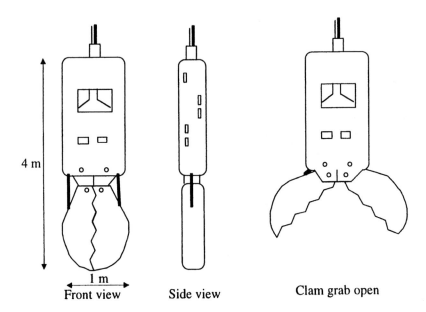

4 m

1 m

Front view          Side view          Clam grab open

Figure 4.20

The main problem of this technique of diaphragm wall construction is obtaining a consistently watertight joint between panels. Since these are formed underground during excavation quality control of the joint is not practical. Figure 4.21 shows the problem.

Figure 4.21

To overcome this problem a steel tube can be positioned at the ends of the excavated panel before concrete is placed, so providing a semicircular joint at the end of the panel, figure 4.22. The diameter of the pipe is the same as the wall thickness and is coated with a debonding agent to reduce the concrete sticking. With two adjoining panels cast the steel tube can be removed and the resulting void grouted.

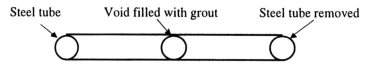

Figure 4.22

Another option is where precast concrete panels have been used. Cast with interlocking joints they can be lowered into the excavated trench by crane instead of concreted in situ, figure 4.23. The resulting joints, however, cannot be made watertight with bentonite and so a replacement cement slurry is used to seal the joint. This cement slurry is a retarded mixture to ensure that it remains fluid during the placement operation but cures sufficiently to grout the joints and does not develop too much strength so that excess can be removed from the front of the wall after excavation to give a regular finish. Again, sealing the joints is difficult and will require some remedial works.

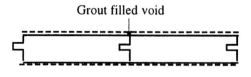

Figure 4.23

The Hydrofraise method of excavation largely overcomes the joint problem because during excavation a portion of the adjoining panel is removed by the revolving cutting head, figure 4.24.

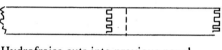

Hydrofraise cuts into previous panel

Figure 4.24

1.5 m

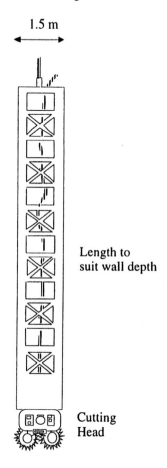

Length to
suit wall depth

Cutting
Head

Figure 4.25

Figure 4.25 shows the Hydrofraise machine, which consists of two rotating cutting heads mounted on the end of a narrow frame of the width of excavation required and length the same as the depth of wall required. The machine can be lowered into the ground by crane and, in operation, excavates the ground, mixes it with bentonite and sucks it to the surface via a tube located at the centre of the frame.

All excavations are continuously filled with bentonite to maintain stability. A reservoir of bentonite is maintained between the guide walls on the surface for this purpose. Panels are cast about 4.5 m long, so the Hydrofraise machine has to be lowered into the ground three times to make a panel; usually two end cuts and then the centre cut. Bentonite is continuously pumped out, filtered and replaced into the reservoir between the guide walls to remove excavated material. Panels are cast alternately, using a tremie tube to place the concrete. Verticality of up to 1 in 500 can be maintained by this system, so the thickness of the wall can be reduced on the conventional excavation system of a clam grab. By adding framed sections to the main body of the machine it can be extended to suit virtually any depth of wall.

### 4.3.3 Slurry Trench Cut-off

This is a cast, in situ cement slurry or concrete wall which is designed to provide a water barrier but not to resist earth pressure, figure 4.26. It can be installed by drilling, augering or any of the excavation methods used for diaphragm walls. It is suitable for silts, sands and gravels and must be used in conjunction with some form of support for the excavation or with a battered excavation.

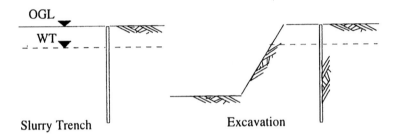

Figure 4.26

### 4.3.4 Thin Grouted Membrane

This is the same as the slurry trench cut off but suitable for poor ground such as silt or a mixture of clay, silt and sand. The grouted void is formed by the penetration of Universal Column steel sections or Rectangular Hollow sections piled into the ground. The sections are removed slowly and the void left by the extracted steel section is grouted under pressure. This is achieved by welding a small diameter tube to the web of the steel section and forcing grout down it under pressure to fill the void before the soil has a chance to close.

### 4.3.5 Contiguous Piling

This is an alternative method to diaphragm wall construction and is a cheaper solution because the rigs used are not specialised equipment and are often used for the installation of piled foundations. The method consists of drilling holes at close centres with an auger of diameter equal to the thickness of the wall. Each hole is augered so that a small gap exists between piles. The size of the gap depends upon the type of soil present in that it is expected to bridge the gap structurally so that the soil is retained. See figure 4.27. This system will not control ground water unless the gaps are filled with some form of water resistant substance such as a water expanding grout. It is desirable to cast a load spreading beam at the top of the piles so that a single pile cannot be overloaded.

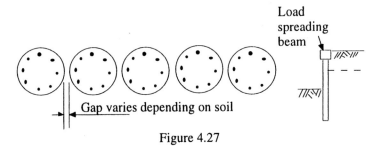

Figure 4.27

## 4.3.6 Secant Piled Walls

This system is often used for the construction of deep basements and can be considered as a cheaper version of a diaphragm wall. It takes the contiguous piled wall one step further towards water control and ground support. The wall consists of a series of interlocking reinforced concrete bored piles, figure 4.28.

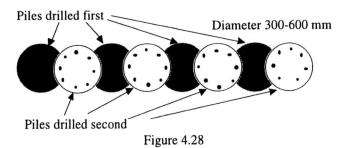

Figure 4.28

Alternate piles are drilled first and then whilst the concrete is still green (set but not hard), the intermediate piles are drilled. The concrete will be a structural mix, say C30, but will have a retarder additive to allow intermediate pile installation. The second piles take the majority of the reinforcement and are called the male piles whilst the first piles installed are called female. The female piles may also contain reinforcement to provide a strong wall to resist lateral loads. Once such a wall is in place the excavated face can be treated with another layer of concrete (sprayed or cast against) to give a fair faced finish. Again a load spreading beam is recommended. See figure 4.29. Such walls are only limited in depth by the capacity of the boring machine and have been constructed to depths of 60 m. Such depths are often stabilised by propping and/or ground anchors.

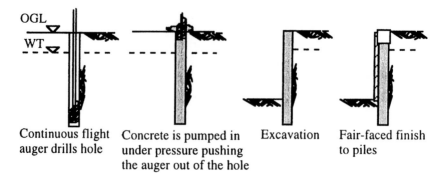

Continuous flight    Concrete is pumped in    Excavation    Fair-faced finish
auger drills hole    under pressure pushing                 to piles
                     the auger out of the hole

Figure 4.29

### 4.3.7 Pressure Grouting

This is used in highly permeable soils such as fissured rocks. Bore holes are drilled at close centres and a cement slurry is pumped in via the drill tube at high pressure. Pressure will build when all fissures and cavities are filled and the drill is slowly removed. This method requires extensive grouting to be sure of obtaining a waterproof curtain and to fill all the fissures and cavities. See figure 4.30. The grout may be composed of a cement, chemical grout, pulverised fuel ash (PFA), or a mixture of all three. PFA is the cheapest to use; it is not as strong as cement grout, and the result is permeable barrier which will not stop water seepage on its own, but it can be used as a bulk filler. It is usual to mix PFA with cement, 10-30 per cent of PFA is common, and this type of mixture can be used as a grout in gravel or a soil with larger voids. A chemical grout can be used in a soil of smaller voids such as sand; these are resin or epoxy mixes and can be 10 times more expensive per cubic metre than a conventional cement based grout.

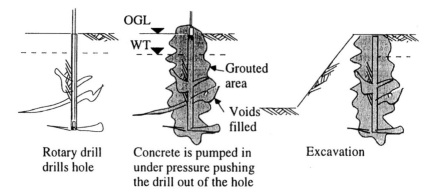

Rotary drill        Concrete is pumped in        Excavation
drills hole         under pressure pushing
                    the drill out of the hole

Figure 4.30

### 4.3.8 Caissons

These structures can be used to construct foundations in open water and bad ground conditions. They consist of a thick concrete box or cylinder between 3 and 15 m in diameter. Open caissons can be used where the ground is clay, but compressed air caissons can be used in any ground. The concrete cylinder is floated into positioned so that the top is safely clear of the water. Excavation is carried out once the inside is pumped clear of water, figure 4.31.

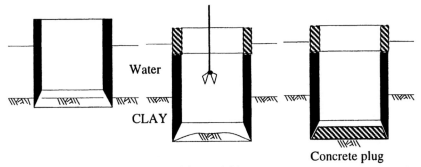

Water

CLAY

Concrete plug

Figure 4.31

As ground is removed the caisson will drop down into the excavation, providing support and water control. Additional rings of concrete wall can be added to the top to keep the water out. Compressed air caissons maintain a positive air pressure at the bottom of the caisson. The air pressure is always maintained at a level higher than the water pressure trying to flood the caisson and so keeps the excavation area free from water whilst the sides of the caisson support the sides of the excavation. See figure 4.32.

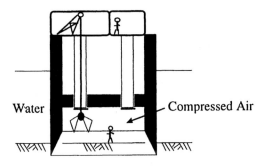

Water          Compressed Air

Figure 4.32

Compressed air working is expensive due to the need for specialised equipment and trained men but it can be effective in otherwise impossible conditions. Compressed air working is also used extensively in tunnelling in poor ground.

## 4.4 OVERVIEW

In most shallow excavations sump pumping will be sufficient, together with some form of ground support. If the ground is made up of sand or silt then a dewatering system will probably be necessary in addition to the above. The design of the support system must be carefully thought out, as this is affected by the type of ground, level of water table and the method of installation. In sand or silt we must also check for the quick condition. If the quick condition is a possibility then sump pumping can be dangerous and induce the condition. If the ground consists of course granular material of high permeability then the quantity of water to be dealt with (called the 'yield') will be a problem and some form of water exclusion system may be necessary in addition to high capacity pumps. In these situations we must be aware of the problems of cavitation caused by the removal of fines from the soil and thus ground settlement. Where settlement can be a problem, for example to nearby buildings, ground water movement must be minimised to maintain equilibrium and some form of cut off trench considered to achieve this. Ground freezing will cause the soil to heave (expand) and will drastically reduce the strength of clay so this method must not be used if ground settlement is a problem or if the soil is later to be used to carry load. Some form of ground improvement may need to be employed in that case.

Figure 4.33 shows a simplified table comparing dewatering systems, water exclusion systems and ground support systems for different soil classifications. We must not forget that a soil profile is very often a mixture of soil types and this is where we must use our engineering judgement to find a suitable solution to ground water control. We will now consider a number of case studies of buildings constructed in recent years.

### 4.4.1 Case Study A - Sizewell B Nuclear Power Station

Positioned on the East coast in Suffolk the geological conditions are 50 m of dense sand and silts known as the Norwich Crag deposits overlying London clay. The problem was that the foundations of the B reactor were designed to be 18 m deep below ground level but ground water levels were recorded as fluctuating at about five metres depth below ground level. Options were either to dewater the ground using deep bored wells, or to exclude the water by using some form of barrier and lower the water level within the barrier.

There were a number of problems associated with deep bored wells. The first was that to lower ground water by 13 m over a 8 hectare site would require a great deal of pumping. Computer modelling suggested that 52 wells would be required, take 12 months to install and become effective, and cost £18m over the five year period that water control was needed. The site is also near the sea so saline water was likely to be drawn into the aquifer and deposit heavy encrustations in the pumps and pipes causing high maintenance costs. Removal of large quantities of water from the ground would cause 'draw down' settlement of the surrounding ground, settlement of the foundations of Sizewell A and drying up of lakes of the bird reserve to the North. It was clear that some form of barrier must be used to limit the effects of the operation on the surrounding

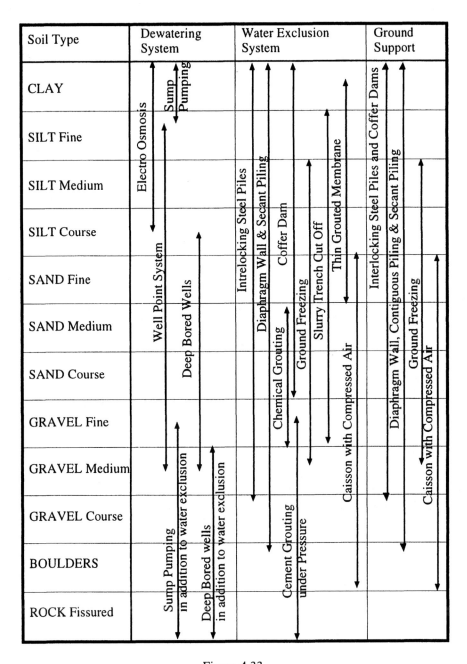

Figure 4.33

environment but here there were also problems. For any barrier to be effective it would have to extend 50 m down to the Oxford clay.

This, together with the requirement to resist a small amount of ground movement and be a structural support for some parts of the work, ruled out the use of ground freezing, grouting or slurry trenches. A diaphragm wall appeared to be the most promising solution but at the time the deepest previous wall was only 30 m deep. The answer came from Europe in the form of a Hydrofraise construction machine from Stent Soletranche. The Hydrofraise machine consists of two rotating cutting heads mounted on the end of a narrow frame 800 mm wide and 56 m deep. The machine could be lowered into the ground and whilst operating would excavate the ground, mix it with bentonite and suck it up to the surface via a tube in the centre of the frame.

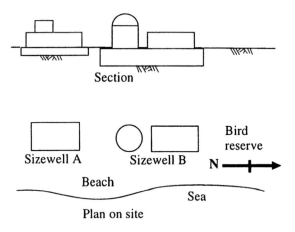

Figure 4.34

All excavations are continuously filled with bentonite to maintain stability. The excavations was filled with concrete via a tremie tube leaving a completed section of wall ready to be added to by the next excavation. The wall was constructed in alternate bays, each bay three excavations long. The bentonite was continually filtered and recycled to remove any remaining debris from the excavation work. The concrete used was a modified retarded plasticised concrete using 50 per cent PFA and designed to withstand 50 mm of lateral movement at the centre without cracking. The result was an all-encompassing cut-off wall 1260 m long and 56 m deep using 54000 m$^3$ of concrete. Nine deep wells were required within the wall which took six months to reduce the ground water to acceptable levels and cost £14m (a saving of £2m).

### 4.4.2 Case Study B - Shoreham Airport

The site location is on the south coast between Brighton and Worthing and is on the west bank of the River Adur and adjacent the sea. The general site is about 20 hectares in size and about 0.5 m AOD. Between the site and the sea is a three metre high earth bund as flood protection during high tides. The ground consists of 0.5 m of topsoil/clay overlying sand/silt six metres thick overlying stiff clay. The water table is one metre below ground level and fluctuates such that the existing grass runways frequently flood especially in the winter. The Client's brief was to provide a hard runway 950 m long by 18 m wide, as shown in figure 4.35, where a grass runway then existed.

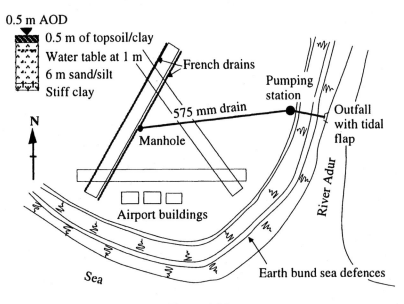

Figure 4.35

The runway had to be drained to ensure continuous use and so a drain of flexible concrete pipe 575 mm diameter was to be installed across the site to a pumping station where the water would be pumped into the river estuary, over the earth bund. The main drainage across the site was between two and four metres deep and discharged into a pumping station was six metres deep.

It was clear that some form of dewatering and support system was required for the excavations. Ground water movements, due to the fluctuation of the tide and the permeable nature of the soil, ruled out the options of a slurry or grouted cut-off wall, and ground freezing was considered to be too slow to cross the existing runways, which were in constant use. A wellpoint dewatering system was chosen because it was suited to the type of ground in question, there were no structures adjacent to the construction area to suffer settlement damage and the depth to which water exclusion was required was within the capacity of the system. Support for the pumping station excavation came in the form of interlocking steel sheet piles around the perimeter, propped by Mabey square brace frames at two metre centres. The sheets were piled down into the soil one metre ahead of, but in pace with, the excavation. The piles had to penetrate the underlying clay to provide ground water control. The drain construction used a well point dewatering system in a rolling mode of operation, and stability was provided by trench boxes (see Chapter 9, section 9.3 for more information on this trench support system).

The drain took five weeks to construct across the site which included five days to cross the runway, during which time it was closed. The construction of the pumping chamber of the pumping station experienced difficulties because after one day of digging the underlying clay was not found, as expected, at 6 m deep. The excavation could not be left until it was stable and work continued through the night. The Contractor considered that there were two options; either continue excavation, adding welded sections to extend the piles, until clay was found, or install a ground freezing system. The Contractor decided to continue excavation and clay was found at 8 m depth. The construction of the main pumping chamber of the pumping station took five days to complete working round the clock. The construction of the rising main through the earth bund sea defences was achieved by constructing a water retaining cofferdam to encompass the outfall construction thus never exposing the airfield site to the risk of flooding.

## 4.5 REFERENCES

1. G. Barnes, *Soil Mechanics,* 1995 Macmillan Press ISBN 0-333-59654-4
2. R Whitlow, *Basic Soil Mechanics,* 3rd Edition 1995, Longman ISBN 0-582-23631-2
3. Williams and Waite, *CIRIA Special Publication 95, The Design and Construction of Sheet Piled Cofferdams,* Thomas Telford, 1993, ISBN 0-7277-1980-7

# 5 Earthworks

All civil engineering work will involve some form of earthworks. This work may be a small or a large part of the construction but, however big, some form of mechanical equipment, called 'Plant' will be employed to carry out the earth-moving operations. In general earthwork can be classified into four broad categories:

• Initial clearance of vegetation and trees and stripping of top soil
• Excavation to form trenches or pits, known as confined excavations
• Cuttings, known as open excavations
• Construction of embankments and filled areas

In this chapter we shall look in some detail at the type of plant used for earthmoving operations, the types of earthworks, the management of earthworks including mass haul diagrams, plant teams and embankment stability.

## 5.1 PLANT

For the earth-moving operation to be economic we must match the right plant to the job. Plant is expensive to buy and operate and must be managed effectively to be economic. Civil Engineers are often responsible for the management of plant and must, at the outset, be completely familiar with the equipment and its capabilities. A suitable choice of plant can only be made after due consideration of the type of soil to be handled, geography of the site, volumes of soil, etc.

Table 5.1 is a summary of the types of plant best suited for different job classifications relating to material. We shall consider in more detail the most commonly used plant in earthmoving operations:

• Motor scraper
• Dozer
• Grader
• Hydraulic excavators
• Dump trucks

A separate section will also consider the safety aspects of working with and around earthmoving equipment.

Table 5.1 Plant summary

| Type of soil | Excavation | Excavate and Load | Haul and Deposit | Excavate, Load, Haul and Deposit |
|---|---|---|---|---|
| Rock | Drill and Blast | Grab | Dump trucks | Dozers |
| Boulders | Breakers | Face shovel | Dumpers | Tractor drawn |
| | Dozers and | Dozers and | Lorries | scrapers |
| | Rippers | Rippers | | Dozers |
| Gravel | Dozers, | Trenchers* | Dump Trucks | Motor scrapers |
| and Sand | Graders | Dragline | Lorries | Dozers |
| | Hydraulic | Loaders | Conveyors | Dredgers |
| | excavators | | | |
| Clay | Trenchers* | Hydraulic | Dump Trucks | Motor scrapers |
| | Hydraulic | excavators | Lorries | Dozers |
| | excavators | (back-hoe ) | Conveyors | Dredgers |

*For trenchers see drainage chapter 9*

### 5.1.1 Motor Scrapers

The motor scraper consists of a large open bucket between two sets of wheels, as shown in figure 5.1. At the bottom of the bucket is a blade which is the width of the machine and can be raised or lowered. The principle of operation is to lower the blade which then cuts into the ground as the machine is moved forward collecting the excavated soil in the bucket. The blade is raised when the bucket is full and the collected soil can be transported to the required location. To unload, the blade is lowered sufficiently to allow the excavated soil to escape whilst the machine moves along and an ejector plate forces the soil forward and outwards in a thin layer ready for compaction. The capacity of the bucket can be anything between 8 and 50 m³.

This machine is designed to move large quantities of soil over considerable distances quickly and is not suited to moving small quantities. Not stopping to load or unload, the cycle is performed on the move at speeds between 5 and 35 km per hour. The operating limitations are that a motor scraper needs a lot of room to manoeuvre and is therefore only suited to bulk open excavations. The machine can only negotiate slight gradients of up to 1 in 30 unassisted and requires the maintenance of a relatively smooth track along which to travel. This is called the 'haul road'.

The maintenance of the haul road is most economically achieved using a grader. Blasted rock or boulders cannot be excavated by motor scrapers because they would get stuck in the gap left by the lowered blade. Indeed the most economic and efficient operation must be considered carefully against soil type. Hard material, such as marl or cemented shale, will be difficult to excavate with the motor scraper and may require a push from a dozer during the loading operation. Sharp rocks may also cause unnecessary wear to tyres (which are very expensive) and cause an uneconomic operation.

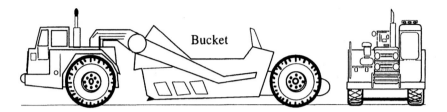

Figure 5.1 Motor scraper

In the case of a motor scraper the machine is self propelled, but the bucket can be mounted between wheels and be drawn along by a tracked dozer; in this situation the bucket is called a box. There are two variations on the standard motor scraper, the first being the inclusion of a second engine powering the rear wheels. This gives the advantage of more traction power to tackle steeper gradients and to excavate harder ground. Of course, it is more expensive to run with two powerful engines, but it is cheaper than running a dozer.

The second variation is called an 'elevating scraper'. This is where a rotating conveyor system assists the excavated soil into the bucket. It has the advantage of speeding the loading and reducing the traction required to load, but it is not often used due to the cost of maintenance against a relatively small advantage.

### 5.1.2 Dozers

Crawler dozers are powerful engines mounted on tracks with a blade fixed on the front. They have high tractive forces which means that they are at home in all soil types and can move up steep gradients very easily. They can move soil by pushing it over short distances (up to 100 m) more efficiently than motor scrapers. Dozers are often used to assist motor scrapers to fill their buckets by pushing them on the filling operation. They are also used to recover construction vehicles which have become stuck or bogged down in the mud.

Dozers can adjust the level and angle of the front blade very accurately which makes them good at grading. See figure 5.2.

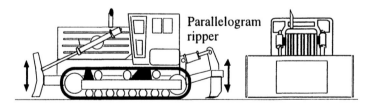

Figure 5.2 Dozer

The compaction action of the tracks is continually at work aiding the grading operation. The front blade has two variations; the first is the straight blade, known as the 'S' or 'Semi U' blade and is shown in figure 5.3. Both blades are used for general dozing and have an extra steel plate welded in the centre to protect the blade from damage when pushing motor scrapers. The second is called the 'U' blade which has extra wings on each side so that it can push more material in one action.

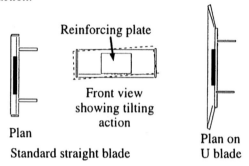

Plan

Standard straight blade

Reinforcing plate

Front view
showing tilting
action

Plan on
U blade

Figure 5.3 Dozer blades

A ripper can be attached to the rear of the dozer which is used to excavate fissured rock, slate or boulders. The ripper is a metal blade known as a shank which is pulled along behind the dozer ploughing the soil up. There are three types of ripper; the first is a radial ripper which is hinged at the fixing point to the dozer and is simply lowered manually into the desired location. The second is the parallelogram ripper which is lowered hydraulically to the required depth and the third is the variable pitch ripper which can be altered hydraulically to adjust the angle of attack. It is possible to rip with one, three or five shanks depending on the nature of the rock and the depth required.

### 5.1.3  Grader

This piece of plant neither excavates nor hauls soil; its job is to grade the finished level of soil to fine limits. It consists of a straight blade, similar to that used on the dozer, suspended under the chassis of a pneumatic tyred vehicle, as shown in figure 5.4. It is used for maintaining the smooth haul road along which the motor scrapers will ride easily. In road construction, the grader can also be used to level large quantities of road materials, such as sub-base, to an accuracy of 10 mm. The pneumatic tyres give the advantage of closing the surface being graded as the machine passes as opposed to tracked vehicles which tend to chop up the surface; but of course pneumatic tyres have less traction.

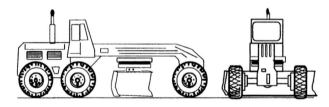

Figure 5.4 Grader

### 5.1.4 Hydraulic Excavators

This plant is arguably the most commonly used equipment today. They are versatile and, with a good driver, can contribute greatly to the general quality of the work in hand. Generally there are two types, the 360° and the 180° machines. The 360° is used for heavy work which may require a long reach of the excavating arm, whilst the 180° is a smaller machine used for more general excavation and loading jobs. Hire costs vary greatly, depending on size of machine and market conditions, but as a guide the hire rate is usually about £15/hr for a 180° and £22/hr for a 360° (at 1996 prices). The 180° machines can travel on the road to site, whereas 360° machines require a low loader which can cost £250 per trip.

*360° Excavators*

The basic principle of operation is a hydraulically operated boom, to which is attached a one or two part dipper stick and bucket. The digging arm is sometimes called a 'back-hoe'. The whole assembly and the driver's cab is mounted on a rotating base which allows excavation and loading to take place in any direction. The base is usually mounted on tracks but can be mounted on wheels. The diagram shown in figure 5.5 is based on the Liebherr R922 Litronic, 100 kW output, operating weight 25 tonne and bucket capacity between 0.3 and 1.7 m³ (shown here with a 0.5 m³ capacity bucket). Such machines can also be fitted with clamshell grabs, pneumatic breakers for breaking rock or concrete and ditching buckets. Ditching buckets are often used for grading or trimming completed excavations because of the wide smooth blade.

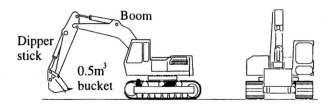

Figure 5.5 360° Excavators

The hydraulic operation gives the arm a high penetration force together with a high degree of precision. The range of operation of the arm is defined by the manufacturer diagrammatically using a sweep area as shown in figure 5.6.

*180° Excavators*

This is the construction version of a large farm tractor. The 180° excavator has a smaller version of the hydraulically operated excavation arm used on the 360°

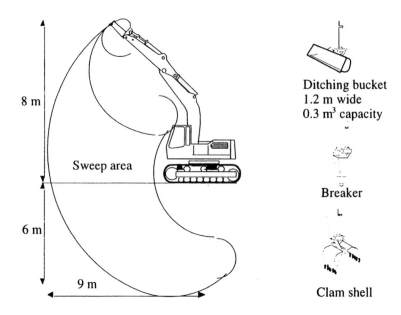

Ditching bucket
1.2 m wide
0.3 m³ capacity

Breaker

Clam shell

Figure 5.6 360° Excavator applications

Liebherr 922 in operation (*Courtesy of Liebherr - Great Britain Ltd.*)

and can take the same accessories. This arm is mounted at the back of the machine and so can only operate within a 180° arc, hence the name. The outline shown in figure 5.7 is based on the JCB 1000B series with its stabiliser feet extended. Weighing nearly 7 tonne it has a wheel base of about two metres, an overall length of eight metres and a height of four metres. The back hoe has a maximum loading reach of two metres and a dig depth of three metres.

A loading shovel mounted on the front of the machine has an approximate capacity of 0.9 m³. This plant is used for small construction work such as digging house foundations and shallow drainage work. It can also be used to load soil using the front shovel. The front shovel shown in figure 5.8 is called a 'three-in-one' and has three operations, to load, grade and dig. Forks can also be fitted to allow fork lifting operation, making this machine a versatile workhorse for the industry.

Figure 5.7 180° Excavator

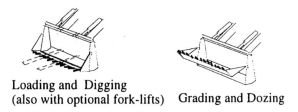

Loading and Digging
(also with optional fork-lifts)    Grading and Dozing

Figure 5.8 180° Excavator applications

### 5.1.5 Dumpers and Dump Trucks

The transportation of soil on site is usually carried out by dump trucks, which are basically off-road lorries and vary in size from 5 to 30 m³ capacity. Such vehicles are heavily reinforced and have robust suspension so that they can handle the rough environment. The outline shown in figure 5.9 is based on a Volvo BM articulated dump truck with 8 m³ capacity. It weighs approximately 18 tonne empty, 25 tonne loaded and can travel at about 50 kph. It has six-wheel drive, each axle with fully independent suspension and is articulated by means of a knuckle-joint in the middle of the main chassis to allow it to travel in rough and wet conditions. Within the author's experience of site operation, it is rare for these types of vehicles to get stuck or bogged down.

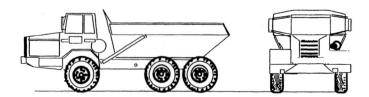

Figure 5.9 Articulated dump truck

### 5.1.6 Safety Aspects of Earthmoving

The plant used for earthmoving is large and powerful and must be treated with the utmost respect. The machinery is noisy and often the driver will only be aware of people on the ground if he can see them and he will naturally be concentrating on the job in hand. All site personnel must do their best to be seen by wearing reflective jackets and hard hats. Obviously, we cannot rely on this alone and so we must remember some important points:

- Keep a safe distance from the machine.
- Never walk close behind a machine or under its boom.
- Always look both ways and listen before crossing a haul road; motor scrapers can move very fast.
- Pedestrians and light traffic should not travel on the haul road.
- Never park on a haul road.
- Generally haul roads should be kept in good condition, by regular grading and compaction as this will reduce braking distances. In the summer dust clouds should be kept to a minimum by regular watering.
- Haul road gradients should be kept to a minimum and large cross falls should be avoided.
- Overhead cables and height restrictions should be clearly marked with signs and tapes.
- It is recommended that all site traffic is fitted with flashing amber lights.

Important aspects of safety and control of large construction plant are overseen by the 'banksman'. A banksman has two parts to his job. The first is to look after the safety aspects of operation and the second is to guide the accuracy of the work. From the safety point of view the banksman is the eyes and ears of the machine driver and he should be present whenever the plant is operating. The banksman will have a ground perspective and be able to see potential dangers before they turn into accidents. For example, he will be able to check the ground to ensure that it is firm, even and free of obstruction and he will look out for situations which may destabilise the machine. The banksman will also be looking out for overhead cables and obstructions and keeping pedestrians and site personnel at a safe distance.

Lorry tipping at the edges of embankments can be a hazardous operation and should always be supervised by a banksman. Tipping should normally take place away from the edge and dozed into position, but if edge tipping is unavoidable then some form of wheel stops must be in position. Generally, plant must not be allowed to work at the top of embankments or close to the edge of excavations as the weight and vibration may contribute towards a slope failure. The stability of an excavated face must be constantly assessed for safety against collapse and if there is any doubt it should be cut back to a shallower slope or some form of support provided. Once the day's work is complete, the banksman will secure any excavations which are to be left open overnight by fencing off with a substantial barrier of some form, if possible lit with hazard lights.

## 5.2 TYPES OF EARTHWORKS

Before we can look in detail at the organisation of plant, we must first consider some common methods of excavation and their application. We shall look at:

- Confined excavation
- Deep or battered excavation
- Open or bulk excavation
- Rock excavation
- Embankment construction
- Grading

### 5.2.1 Confined Excavation

Trenches and foundations are called confined excavations and plant teams of a 360° excavator (back-hoe) and lorry or dump truck are typically used for such work, see figure 5.10.

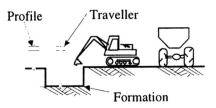

Figure 5.10 Confined excavation

Back-hoe mounted breakers are more likely to be used in situations of this sort, since even soft rock can be difficult to excavate when confined. Loading time is the term used to define how long it takes the 360° excavator to load a lorry and the cycle time is the time it takes the lorry to receive its load, travel to tip, tip, and travel back again. Loading times in general would be about 10 minutes and so the cycle time for the lorry (or dump truck) is in general about 20 minutes, depending upon how far away the tip is. We can thus see that in general two lorries are required for each 360° excavator and are regarded as a basic plant team.

The setting out of such an excavation is accomplished using sand lines to mark the edges and profile board and traveller to define the depth to formation, as shown in figure 5.10. A profile board is a short length of wood nailed to a post at the side of the excavation. The level at the top of the board is set (usually at two metres) above the intended bottom of the excavation known as the formation. A traveller is a post two metres long with a short board nailed to the top. Usually, two profile boards are erected either side of the excavation and the banksman looks across the two boards whilst the traveller is held in the excavation. The excavation is at the right depth when the profiles line up with the top of the traveller.

### 5.2.2 Battered Excavations

Battered excavations are required if the excavation is deep and the soil needs to be stabilised against collapse using a slope. Generally, plant teams are the same as for confined excavation but the work takes place in two stages as shown in figure 5.11.

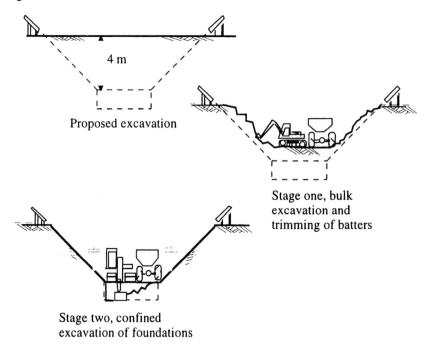

4 m

Proposed excavation

Stage one, bulk
excavation and
trimming of batters

Stage two, confined
excavation of foundations

Figure 5.11 Battered excavation

The first stage of the excavation is that of the slopes and continues down until we are at the same level as the top of the confined excavation required. The second stage excavation, which is essentially a confined excavation, can now begin.

A batter is the term used to describe the slope of the sides of the excavation, and this type is called an 'open cast' excavation. It is usual to describe the batter as, for instance, a '1-in-2 slope', meaning that the slope rises in height by one metre for every two metres travelled horizontally. Such slopes are designed to be stable for the soil in which they are excavated, so the slope will vary depending upon the type of soil under consideration, see section 5.4 on slope stability. The advantage of this method is that it allows unrestricted access to the excavation and is cheaper than supporting the sides of the excavation with piles and props. Its disadvantages are that the excavation is bigger, requiring more land and the volume of soil to be removed is greater. Obviously, some form of ramp must also be excavated to allow access by machines.

Here the setting out, as the excavation, takes place in two stages. First, batter rails are set up at the same angle as is required for the slope (batter), these point to the ground at the intended start of excavation. A sand line may also be used to define the extent of the excavation. The batter rails have a slope distance written on their side and this is measured by the banksman to ensure that the excavation is at the correct depth. During the excavation the banksman sights down the top of the rail to ensure that the correct slope is being dug. Setting out of the confined dig can, once again, be defined with sand lines to show the extent of the excavation, and with profile boards and a traveller to define the depth.

The setting out of the batter boards can be a tricky operation, especially if the original ground level is part of a slope. The following is a simple iterative method for locating the profile position relative to the centre line of the excavation. In figure 5.12, the height of A above D is $(H + h - H_1)$ and the height of C above D is $(H + h - H_2)$, then the following equations apply.

$$w_1 = \frac{b}{2} + m(H + h - H_1) \quad \text{and} \quad w_2 = \frac{b}{2} + m(H + h - H_2)$$

With the level set up as shown in figure 5.12 take a staff reading at the centre line B. Read the staff at a trial position of A, whilst taping horizontally from B. Calculate the value of $w_2$ from the above equation and, if it agrees with the taped reading, then the position is correct.

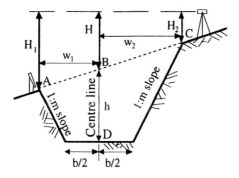

Figure 5.12 Setting out batters

### 5.2.3 Bulk Excavation

Generally larger plant can be used for bulk or open excavation because there is more room to manoeuvre and larger quantities to be move. A plant team may consist of 2 or 3 motor scrapers, a grader to maintain the haul roads and a dozer to assist the motor scrapers and grade the batter. A 360° excavator is sometimes used to finish the slope and spread topsoil. The excavation is generally open but must be carried out in two stages, the first stage is as shown in figure 5.14 the second in figure 5.15. Again the setting out, as for the excavation, is carried out in two stages, first the erection of batter boards to define the top and slope of the intended batter, then the second stage is to put up profile boards to define the formation level.

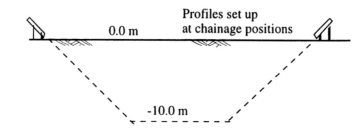

Figure 5.13 Proposed excavation

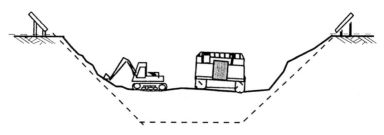

Figure 5.14 First stage excavation

Figure 5.15 Second stage excavation

Some batters are inaccessible to tracked vehicles, either due to weather or the steepness of the slope, so trimming of the batter must be carried out by a 360° excavator standing at the top or the bottom of the slope. Obviously, the slope cannot be too long for the reach of the 360° machine and so bulk excavation may have to be halted to allow this operation.

### 5.2.4 Rock Excavation

Strength of rock varies with type and location. Rocks which are relatively easy to excavate in the open using a ripper are difficult to excavate in confined spaces. Weak rock (silt stone or mud stone) can be excavated with standard plant. Hard rock (such as slate and sandstone) can be ripped or 'gunned out' using a pneumatic breaker. Very hard rock (such as limestone and granite) may have to be blasted.

There are a number of methods used for breaking very hard rock in addition to blasting. Obviously, pneumatic breakers can be used, but these are slow and expensive and are usually used for small excavations and for breaking large boulders. Drilling can be used to break rock by driving plugs into the resulting holes or freezing liquid in the holes so that it expands. Both of these methods work by applying mechanical pressure from the inside thus cracking the rock.

The most commonly used method for rock excavation is drill and blasting. This technique involves drilling a grid of holes at about one metre centres throughout the area to be excavated. The holes are charged with small capsules of gelignite, 'stemmed' (backfilled) with sand or clay and detonated. This reduces the solid rock to rubble which can then be excavated by conventional plant. The diameter of the drilled holes is important because the larger the diameter the more charge is required thus increasing noise and vibration. In general drilled holes of 38 mm diameter are used for depths of up to 3.5 m, and between 50 mm and 75 mm diameter for depths of 3.5 to 9 m. Blasting usually takes place using the benching method and holes are usually drilled to a depth 0.5 m below the desired excavation depth, see figure 5.16.

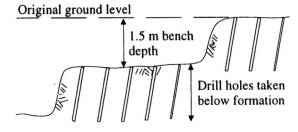

Original ground level

1.5 m bench depth

Drill holes taken below formation

Figure 5.16 Benching method

Open cast gelignite is usually used for this type of work. It is a medium strength explosive which is reliable in wet conditions and fragments the rock well. Different types and amounts of explosives should be matched to the hardness of the rock, the size of the pre-drilled hole, the depth of excavation and the spacing of the holes and an expert should be consulted. In general the amount of explosive used varies from 250 g/m³ in soft rock to 500 g/m³ in hard rock. The benching method is used for the primary blast and if the boulders produced are too large to handle then secondary blasting is used on individual rocks.

In general blasting can leave an irregular excavated surface which, if it is to form part of the finished surface of a slope, is undesirable. In this situation a technique called pre-blasting is used. This method involves drilling holes at one metre centres along the same plane as the desired finished slope. These holes are charged and stemmed in the usual way but are detonated a few milliseconds before the rest of the charges. The shock of the blast is neutralised along this plane producing a reasonable finished surface.

### 5.2.5 Embankment Construction

This is usually a bulk operation and is carried out in much the same way as excavation. Again, we require batter rails to define the intended slope; the only difference this time is that a traveller is used, usually one metre high, as shown in figure 5.17. Setting out is similar to the process given in section 5.2.2.

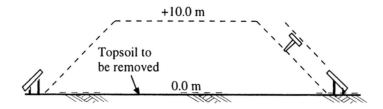

Figure 5.17 Proposed embankment

Compaction and grading plant will work continuously with the haul operation, but the rate of construction must be controlled so that the increase in loading will not over-stress the subgrade/subsoil. Constant checks are made of pore-water pressure in the subsoil throughout construction. Once again batters may be inaccessible to a dozer and so trimming must then be carried out by a 360° excavator. Obviously, the slope must not be allowed to become too long for the reach of the 360° machine and so the rate of embankment construction may have to be adjusted to allow for trimming. See figures 5.18 and 5.19.

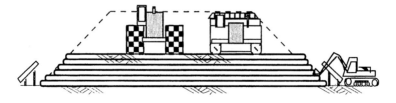

Figure 5.18 First stage fill and compaction

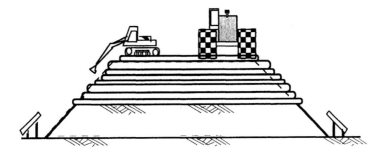

Figure 5.19 Second stage fill and compaction

### 5.2.6 Grading

Once the second stage is complete in either excavation or embankment construction, the final levelling is carried out by a grader. Lasers are often used in the grading operation to achieve quick and accurate results. The system works by erecting a laser level in the centre or to one side of the area to be graded. A laser detector is attached to the blade of the machine and a diagrammatic readout given to the driver via a liquid crystal display mounted in the cab. The display simply shows when the blade is too high or too low and is adjusted by the driver. More recent systems can link the blade directly to the laser detector so that any adjustments take place automatically.

The exact level used to define the final grade must be considered carefully. In the case of excavation (if the soil is clay or has a clay content) then when the overburden is removed heave occurs. This means that the soil at formation level will rise over a period of time after the excavation. The reason for this is that the soil just below the formation was then compressed by the weight of the soil above, but if this is now removed then a decompression must occur allowing an increase in volume. This process takes between 6 months to one year and final grading can only be carried out when it is complete. A layer of soil is often left in place over the formation during winter as protection from weather and plant movement and removed by final grading. The protection layer should be graded to falls to prevent the accumulation of water and the resultant softening of the formation. Certain tolerances are acceptable in road construction, for instance the DoT allows +20 to −30 mm. The Engineer must take all of the above factors into account when defining the final grade level.

In the case of embankment construction two types of settlement must be considered. The soil with which the embankment is constructed will settle within itself both during construction and after. The underlying soil will be compressed and settle due to the extra weight called 'overburden'. For these reasons the level of an embankment is left high until the effects have equalised and then final grading is carried out. This is called consolidation and can take up to 12 months. Again a cross fall should always be graded onto the formation so that rainwater may drain away freely. The weather may have a considerable effect on the surface soil which if softened will have to be removed before the placement of the road construction.

## 5.3 MANAGEMENT OF EARTHWORKS

The plant that we have been considering is expensive to buy and operate and so we must ensure that it is run efficiently in today's competitive markets. There are many factors which must be taken into consideration in controlling the planning and execution of earthworks. The main factors to take into account are:

- Type of material
- Geographical factors - access, size of site and weather
- Volume of soil to be moved and the time restraints
- Plant available
- Legal restraints on noise, disturbance, etc.

The type of material to be moved is very important, both in terms of physical properties and assessing likely costs. For example, topsoil is easy to excavate but it is often reused and has to be kept separate from the rest of the excavation. This requires double handling, so for this reason it is billed at a higher rate than the general excavation; £5.00-£10.00/m³ of topsoil, as opposed to £2.50 - £5.00/m³ for general excavation, at 1996 prices. Clay is an easy material to excavate but is difficult to travel over and compact in wet weather and it can readily become unsuitable as a filling material if it should become too wet. The hardest material to excavate is rock which may need specialist equipment such as hydraulic breakers or explosives; this together with its high bulking factor, makes it an expensive material to handle.

All materials occupy a greater volume when excavated than they did before excavation. This phenomenon is known as bulking and results from the inclusion of a greater volume of air voids within the material due to the disturbance of excavation. The volume of the soil before excavation is called the 'solid' volume, or sometimes the 'banked' volume, and the volume of soil after excavation is called the 'bulked' volume. Bulking factors for different materials are given in table 5.3. After compaction the soil will occupy a greater volume than the solid volume. This is a common feature of all material except chalk which actually occupies less volume when recompacted than it did when in the solid state. This is due to the fragile structure of the soil. If for any reason there is a short-fall of fill material, extra material can be excavated on site if available; these excavated sites are known as borrow pits.

Geographical factors such as the location, size and the shape of a site are also important. The location of the site will give a good indication of the type of weather to expect, which is an important factor in controlling trafficability. Rainfall tends to soften the ground and make access difficult. It also increases moisture content in compacted fill, making it difficult to compact. The topography of a site such as size, terrain and access can mean the difference between dump trucks and motor scrapers. There are operating limits of gradients for motor scrapers which are well within the capabilities of dump trucks.

Volumes of soil mass and the movement details are factors which must be considered for effective earth-moving management. One tool used to plan and represent this complex task is the mass haul diagram, as described in section

5.3.1. Once the type of material, geographical factors and the volume have been assessed, the output rates of plant teams can be matched to the task in hand. Output information can be gained from the manufacturer or by comparison with a previous task. This is also looked at in more detail later.

Contractual and legal restraints, such as a restriction on working hours or noise outputs, may be controlled by statutory regulations, such as those in BS 5228 'Noise Control on Construction and Open Sites', or special provision under the contract to minimise the effect of the works on the local inhabitants. Preservation orders on trees or buildings, mass trespass by protesters, and even badger sets, may have a potential effect on the work.

### 5.3.1 Mass Haul Diagram

For an economic operation it is essential to match the right plant to the right job. To do this we must look at the task in hand carefully and assess:

- Volume of material
- Distance of haulage
- Duration or timing of the work

In larger excavations the effects of volume, distance and timing are assessed using a mass haul diagram. This is a graphical representation of the volume of cut and fill in relation to its position on the site and is typically used for road construction. The distance along the centre line of a road is called the 'chainage' and is measured in metres from an agreed zero point called 'zero chainage'. The two sides of the centre line are defined as 'left and right hand channel' and are always defined with one's back toward zero chainage.

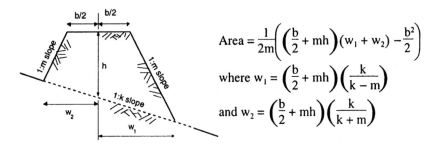

$$\text{Area} = \frac{1}{2m}\left( \left(\frac{b}{2} + mh\right)(w_1 + w_2) - \frac{b^2}{2}\right)$$

$$\text{where } w_1 = \left(\frac{b}{2} + mh\right)\left(\frac{k}{k-m}\right)$$

$$\text{and } w_2 = \left(\frac{b}{2} + mh\right)\left(\frac{k}{k+m}\right)$$

Figure 5.20

To calculate the volume of earthworks we consider a longitudinal section showing the proposed road and existing ground levels. By considering the difference between the two, the areas of cut and fill can be calculated at each chainage position and the volumes found. A simple equation for calculating the area of a cross-section is shown in figure 5.20. The same equation can be used for cuttings by simply turning the diagram upside-down. This can be a lengthy procedure but is simple mathematics and is ideal for computer application.

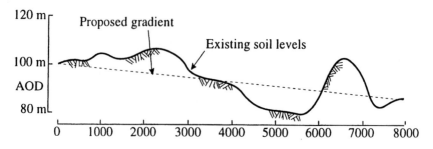

Figure 5.21 Longitudinal section

Consider the example shown in figure 5.21. This is a typical longitudinal
section of a proposed road and has an orthogonal scale of 20 m to 12 mm
vertically and 1000 m to 12 mm horizontally. The volumes of cut and fill have
been worked out and are presented in table 5.2. Once the volumes are known we
can show them diagrammatically in the form of a mass profile diagram as in
figure 5.22.

Table 5.2 Cut and fill quantities

| Chainage(m) | Volume in m³ | | Cumulative |
| | Cut | fill | |
|---|---|---|---|
| 0 - 1000 | 10,000 | | 10,000 |
| 1000 - 2000 | 20,000 | | 30,000 |
| 2000 - 3000 | 30,000 | | 60,000 |
| 3000 - 4000 | - | | 60,000 |
| 4000 - 5000 | | 30,000 | 30,000 |
| 5000 - 6000 | | 40,000 | −10,000 |
| 6000 - 7000 | 40,000 | | 30,000 |
| 7000 - 8000 | | 10,000 | 20,000 |

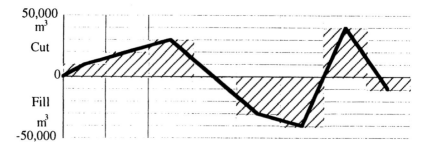

Figure 5.22 Mass profile

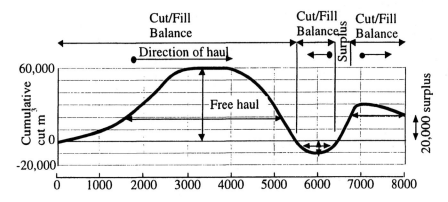

Figure 5.23 Mass haul diagram

The mass haul diagram in figure 5.23 shows the cumulative volumes of excavation beginning at zero chainage on the left and moving right. From the mass haul diagram we can obtain much information to help manage the task of earth-moving. The first information apparent is the number of cut/fill balance areas and the length of chainage for each. This indicates the location of each individual 'muck shift' operation and its relative size. Also the direction of haul is indicated as left to right for a peak and right to left for a trough. We can also see that between chainage 0 and 6000 there is sufficient bulk to warrant using motor scrapers. Between chainage 6000 and 8000 the gradients of terrain are steeper than elsewhere, so 360° excavators and dump trucks will be necessary initially. We can see that the muck shift operations are self sufficient from chainage zero to chainage 6500 and we have to remove a surplus of 20,000 m³ from between chainages 6500 and 8000. This we can assume will be taken off site to a tip, so road lorries will be needed and an access suitable for them constructed at that location. So we can see a strategy forming, i.e.

• use motor scrapers chainage zero - chainage 6000
• use 360° excavator plus dump trucks chainage 6000 - chainage 8000
• lorries will require access between chainage 6000 - chainage 8000.

From the costing point of view we can see the total amount of soil to be moved and the average distance that the plant will have to travel throughout each muck shift operation, known as the 'free haul'. This information is found from the vertical ordinate and abscissa which intersect at the centre of gravity of each cut/fill balance area. For this example the free haul is 3.5 km between chainages 1700 and 5200. The total volume of muck shift over this distance is 60,000 m³. The free haul is not really free but is simply the bench-mark used when a contract is in progress to describe earthworks carried out at bill rates. Any variations to the average distance of travel may be subject to a claim for extra money by the Contractor.

When the mass haul diagram is drawn we can experiment with the 'what-if' scenarios to establish the most efficient method of working. In the example considered so far, the 'line of balance' is the zero volume line, but this can be moved up or down to adjust the size and length of each muck shift operation. For example it may be moved up to the 20,000 line as shown in figure 5.24. This would indicate that the muck shift is self sufficient from chainage 8000 down to 1500 and that a surplus now occurs between chainage 0 and 1500. Now we have four different operations, two between chainage 1500 and chainage 6500, of 40,000 m³ and 30,000 m³ which could use two teams of motor scrapers. Between chainage 6500 and chainage 8000 there is a 10,000 m³ operation with one team consisting of a 360° excavator and some dump trucks to cope with the gradients. The surplus can be removed by 360° excavators and road lorries at chainage zero. Once we have a clear idea of the operations required, we can look in some detail at the composition of the plant teams necessary to complete the job on time.

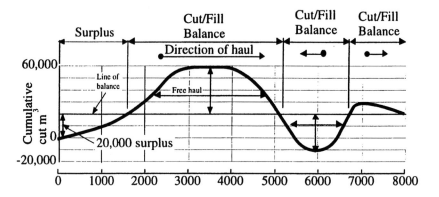

Figure 5.24 Adjusted mass haul diagram

### 5.3.2 Plant Teams

The composition of a plant team is a critical decision which will affect the efficiency of work on site and needs to be made by management before work commences. The process must draw together all the factors considered so far to enable a calculation to be carried out which will predict the rate of working and thus profitability. These calculations are best explained by example.

Consider the 20,000 m³ that goes to tip in the above example. The material is clay and so a bulking factor of 1.3 will be applied. (See table 5.3 for a range of bulking factors.) Work will be carried out in early summer and we have 10 weeks programmed for it.

$$\text{loose volume} = 1.3 \times 20,000 \text{ m}^3 = 26000 \text{ m}^3$$

As already stated, there is a range of bucket sizes from different manufacturers from 0.3 to 2.5 m³ capacity. The capacity of a bucket is usually based upon heaped capacity, but struck capacity is sometimes used, see figure 5.25.

Table 5.3 Bulk density and bulking factors

| Soil type | Bulk density in situ kN/m³ | Bulking factor |
|---|---|---|
| CLAY | | |
| soft | 1.60 - 1.95 | |
| firm | 1.75 - 2.10 | 1.20 - 1.40 |
| stiff | 1.80 - 2.25 | |
| Topsoil | 1.35 - 1.40 | 1.25 - 1.45 |
| Peat | 1.05 - 1.40 | 1.25 - 1.45 |
| SAND | | |
| fine | 1.60 - 2.10 | |
| medium | 1.75 - 2.20 | 1.10 - 1.15 |
| course | 1.90 - 2.25 | |
| Gravel | 1.70 - 2.25 | 1.10 - 1.15 |
| Chalk | 1.65 - 2.40 | 1.30 - 1.40 |
| Schist and slate | 2.70 - 2.90 | 1.30 - 1.65 |
| Sandstone | 2.46 - 2.65 | 1.40 - 1.70 |
| Limestone | 2.40 - 2.70 | 1.45 - 1.75 |
| Granite | 2.60 - 2.70 | 1.50 - 1.80 |

Heaped    Struck

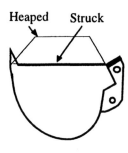

Figure 5.25

The bucket fill factor is a proportion of the bucket heaped capacity which is applicable to different types of soil. Sometimes the soil may be too bulky to fill the bucket (such as large boulders). See table 5.4 for typical bucket fill factors. We shall consider using a 360° excavator with a bucket of heaped capacity 0.75 m³. Load per cycle = 0.8 (bucket fill factor) × 0.75 = 0.6 m³.

The 'cycle' is the word used to describe the digging cycle, i.e. to fill the bucket, position the bucket over the deposit site, such as a lorry, release the load from the bucket and reposition the bucket to dig again.

Table 5.4 Bucket fill factors

| Material | Bucket fill factor |
|---|---|
| Moist or Sandy Clay | 1.00 - 1.10 |
| Sand and Gravel | 0.95 - 1.10 |
| Stiff Clay | 0.80 - 0.90 |
| Boulders | 0.60 - 0.75 |
| Poorly Blasted Rock | 0.40 - 0.50 |

The number of cycles required to fill a 14 m³ lorry (20 tonne) = 14/0.6 = 24
A typical cycle time is half a minute but this can vary depending upon the digging force and the type of soil and it is best to consult the manufacturer's information on the particular machine under consideration.
Load time per lorry = 24 × 0.5 = 12 minutes, plus 2 minutes allowed for the lorry to position itself under the 360° excavator, called 'positioning time'.

LOAD TIME = 14 minutes

We can now calculate the capacity of the 360° excavator, assuming a normal working week of 40 hours. In doing this we must also consider that some time is spent on refuelling and servicing, known as 'down time'. In order to include down time we will allow a typical figure of working 50 minutes in every hour, which is 83 per cent efficient. Again this is a typical figure and it is best to consult the manufacturer's information of the particular machine under consideration. Thus for a 360° excavator for a 40 hour week working 50 min/hr:

$$\text{Output} = \frac{40 \times 50}{14 \ (minutes)} = 143 \text{ lorries/week}$$

$$= 143 \times 14 \ (capacity) = 2000 \text{ m}^3/\text{week}$$

Now consider the lorry capacity. We know that it can carry 14 m³ but we also need to know how long it will take to transport the soil to tip and come back again ready for the next load. This is called the 'work cycle'.

| Lorry work cycle is: | loading | 14 (from above) |
|---|---|---|
| | moving off site | 1 |
| | travel 2.5 km (say) | |
| | each way (at 30 km/hr) | 10 |
| | tipping time | 3 |
| | | Total 28 minutes |

$$\text{Number of lorries required} = \frac{28 \ (minutes)}{14 \ (minutes)} = 2$$

Load time is therefore a controlling factor. From the programme we have 10 weeks to complete 26,000 m³ as loose dig (20,000 m³ as solid dig). Therefore one 360° excavator is insufficient.

$$\text{Time taken} = \frac{26000}{2000} = 13 \text{ weeks}$$

From the above calculations we can see that a practical plant team for the removal of 20,000 m³ of soil to tip will take 13 weeks and consist of:

One 360° excavator
Two 14 m³ lorries

To meet the programme we can either hire a second similar plant team for the last 3 weeks of the programmed period, have two plant teams working for 6 weeks then one to finish off, or have one plant team on 'bonus'! We can now calculate the costs for 13 week completion and submit the price to the Client. If we decide to use a bigger excavator and or lorries this too can be costed and an economic decision made as how best to tackle the job.

Lorries £19.50/hr × 2 × 40 × 13   = 20,280
Hymac £20.78/hr × 1 × 40 × 13   = 10,806
Banksman £5.11 × 2 × 40 × 13    = <u>5,314</u>
                    Total   £36,400

This works out at £1.82 /m³ of solid dig not allowing for cleaning the roads, tip fees, overheads or profit.

### 5.3.3 Compaction Theory

Once we have calculated the costs and capacities of excavation we also have to consider the compaction activity. This is because often most of the soil excavated is compacted in another location and the compaction activity, if slower, can lengthen the whole operation. Nevertheless, we must be able to manage the plant efficiently to compete on price and to do that we need to understand the physics of the compaction process. Compaction is the process where particles of soil are mechanically packed together to reduce the air content. Figure 5.26 shows the typical soil model for the compaction process.

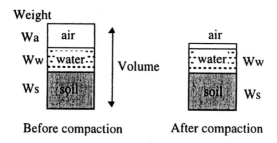

Figure 5.26 Soil model

One way of measuring the compactness or density of a soil is to measure the weight of the dried soil and look at it as a ratio to the volume of the soil in situ. This is called the 'dry density'. Moisture content is also very important when considering the compaction behaviour of different soils.

$$\text{Dry density} = g_d = \frac{\text{Weight of the soil } W_s}{\text{Volume } V}$$

$$\text{Moisture content } m = \frac{\text{Weight of the water } W_w}{W_s} \times 100\%$$

First water acts as a lubricant to achieve greater compaction. Then as the air content decreases, the water tends to keep soil particles apart and so hinders compaction. We can show this graphically if a soil is subjected to the same compaction force and compacted a number of times for a varying water content. We can see from figure 5.27 that the dry density will increase to a maximum value and then begin to decrease. The maximum dry density will occur at a single moisture content, and this is known as the optimum moisture content.

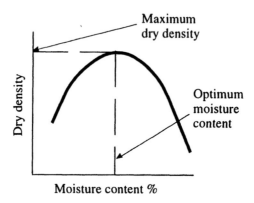

Figure 5.27

The significance of this behaviour is that a soil cannot be fully compacted unless the moisture content is at the optimum value. In dry weather water will often have to be added to the soil to achieve the correct compaction; conversely compaction cannot continue in wet weather and often moisture content readings are taken of compacted soil to ensure full compaction.

Dry density or the amount of compaction can be tested by a number of methods.

- Core cutter method (BS 1377[2]:1990: test 15D)
- Sand replacement method (BS 1377[2]:1990: test 15A to C)
- Radiation

The most commonly used method these days is by radiation, see figure 5.28. Here a low energy radiation source is inserted into the soil to a depth of about 150 mm and a detector on the surface positioned adjacent to the source detects the radiation which travels through the soil. The more dense the compaction, the less radiation is received by the detector and a direct digital reading is given by the detector of the dry density of the soil. This equipment is very easy to use but the main problem is the difficulty of storage of the machine. Storage must be secure, shielded from all personnel and licensed.

The sand replacement test, figure 5.28, uses a sand of known density to fill a hole cut in the soil under test. The weight of sand used is carefully measured and so the volume of the cut hole can be calculated using the equation density = mass/volume. The soil dug from the hole is dried and weighed giving the dry weight and so the dry density can be found.

The core cutter method is seldom used but involves cutting a core from the material under test. The sample is sealed with wax and sent back to the laboratory for tests.

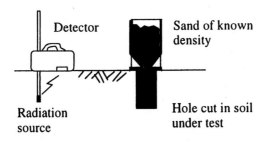

Radiation test          Sand replacement test

Figure 5.28

### 5.3.4 Compaction Plant

Once the soil has been transported to its destination, it then has to be spread and compacted. (If the soil is transported by motor scraper, however, the soil is spread by the way that the soil is released from the machine and then only compaction is required.) There are three main compaction actions used by rollers: dead weight, kneading and vibration. The action chosen is that which is best suited to the soil to be compacted and indeed a combination of actions is often used. Clay does not respond well to compaction by vibration but it compacts efficiently under the dual action of dead weight and kneading. Granular material on the other hand responds well to compaction under dead weight and vibratory action. See figure 5.29 for a summary of compaction actions for different soil types.

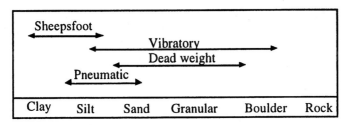

Figure 5.29

When dealing with soft rock, such as chalk, a special watch must be kept on the moisture content because it will degrade very quickly in wet conditions due to the unstable soil structure. Earthmoving operations in chalk therefore tend to be carried out with excavator and dump trucks rather than motor scrapers. Compaction of bulk fill is generally carried out by towed or motorised vibratory rollers. The towed roller is usually pulled by a dozer, and in so doing the dozer spreads and compacts fill in one pass. See figure 5.30. The roller weighs about 7 tonne and has within the drum an eccentric weight rotated by a diesel engine. This has the effect of vibrating the drum and the soil over which it passes giving a greater compacting action.

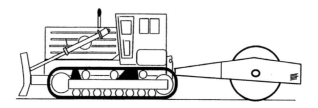

Figure 5.30

Smooth-wheeled vibrating rollers are often built as self propelled plant such as that shown in figure 5.31. This outline is based on the Bomag BW 212-2 which has an operating weight of approximately 9 tonne, a roller width of 2100 mm and a maximum speed of 18 km/h.

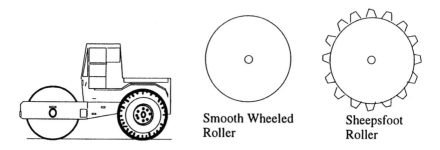

Smooth Wheeled
Roller

Sheepsfoot
Roller

Figure 5.31                         Figure 5.32

The drum of the roller is smooth but can also have on its surface rows of protruding steel feet, figure 5.32. The feet knead the soil as the roller passes giving a greater compaction effect. This type of roller is particularly useful for compacting clay or a soil with a high clay content. Such a roller is called a 'pad foot' or 'sheepsfoot' roller. The outline shown in figure 5.33 is based on the Bomag BC601 RB Land Fill Compactor which has an operating weight of approximately 28 tonne and a blade width of 3.8 m.

## Table 5.5 Compaction specification

| Category | Cohesive soil | | Well graded granular and dry cohesive soils | | Uniformly graded material | |
|---|---|---|---|---|---|---|
| | D | N | D | N | D | N |
| **SMOOTH WHEELED ROLLER** Mass per metre width of roller | | | | | | |
| 2100 kg - 2700 kg | 125 | 8 | 125 | 10 | 125 | 10* |
| 2700 kg - 5400 kg | 125 | 6 | 125 | 8 | 125 | 8* |
| over 5400 kg | 150 | 4 | 150 | 8 | U | - |
| **GRID ROLLER (Sheepsfoot) Mass per metre width of roller** | | | | | | |
| 2700 kg - 5400 kg | 150 | 10 | U | - | 150 | 10 |
| 5400 kg - 8000 kg | 150 | 8 | 125 | 12 | U | - |
| over 8000 kg | 150 | 4 | 150 | 12 | U | - |
| **TAMPING ROLLER Mass per metre width of roller** | | | | | | |
| over 4000 kg | 225 | 4 | 150 | 12 | 250 | 4 |
| **PNEUMATIC-TYRED ROLLERS** Mass per wheel | | | | | | |
| 1000 kg - 1500 kg | 125 | 6 | U | - | 150 | 10* |
| 1500 kg - 2000 kg | 150 | 5 | U | - | U | - |
| 2000 kg - 2500 kg | 175 | 4 | 125 | 12 | U | - |
| 2500 kg - 4000 kg | 225 | 4 | 125 | 10 | U | - |
| 4000 kg - 6000 kg | 300 | 4 | 125 | 10 | U | - |
| 6000 kg - 8000 kg | 350 | 4 | 150 | 8 | U | - |
| 8000 kg - 12000 kg | 400 | 4 | 150 | 8 | U | - |
| over 12000 kg | 450 | 4 | 175 | 6 | U | - |
| **VIBRATING ROLLER Mass per metre width of vibrating roller** | | | | | | |
| 270 kg - 450 kg | U | - | 75 | 16 | 150 | 16 |
| 450 kg - 700 kg | U | - | 75 | 12 | 150 | 12 |
| 700 kg - 1300 kg | 100 | 12 | 125 | 12 | 150 | 6 |
| 1300 kg - 1800 kg | 125 | 8 | 150 | 8 | 200 | 10* |
| 1800 kg - 2300 kg | 150 | 4 | 150 | 4 | 225 | 12* |
| 2300 kg - 2900 kg | 175 | 4 | 175 | 4 | 250 | 10* |
| 2900 kg - 3600 kg | 200 | 4 | 200 | 4 | 275 | 8* |
| 3600 kg - 4300 kg | 225 | 4 | 225 | 4 | 300 | 8* |
| 4300 kg - 5000 kg | 250 | 4 | 250 | 4 | 300 | 6* |
| over 5000 kg | 275 | 4 | 275 | 4 | 300 | 4* |
| **VIBRATING PLATE COMPACTOR Mass per unit area of base plate** | | | | | | |
| 880 kg - 1100 kg | U | - | U | - | 75 | 6 |
| 1100 kg - 1200 kg | U | - | 75 | 10 | 100 | 6 |
| 1200 kg - 1400 kg | U | - | 75 | 6 | 150 | 6 |
| 1400 kg - 1800 kg | 100 | 6 | 125 | 6 | 150 | 4 |
| 1800 kg - 2100 kg | 150 | 6 | 150 | 5 | 200 | 4 |
| over 2100 kg | 200 | 6 | 200 | 5 | 250 | 4 |

*D = Maximum depth of compaction layer in mm*
*N = Minimum number of passes,  U = Unsuitable*
*\*Note roller must be towed by a track laying tractor*
*Crown copyright. Reproduced with the permission of the Controller of HMSO*

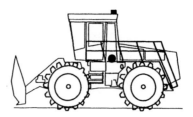

Figure 5.33

Pneumatic rollers are simply rollers on tyres instead of a steel drum. This type of roller is best suited to granular materials consisting of fine particles such as sand or silt. The moisture content must be near its optimum (see later) in such material for it to compact correctly, but the pneumatic action tends to close the surface, whereas the soil tends to stick to a steel drum. It is also used for road resurfacing work because it will not damage the road surface.

The minimum number of passes for each roller and the thickness of compacted layers are shown in table 5.5 and are specified by the DoT in the Specification for Highway Works[3].

### 5.3.5 Compaction Rate of Working

The selection of compaction plant is important to the overall cost-effectiveness of the operation. The choice of plant depends upon material factors and spatial factors. The plant must be suitable for the material to be compacted and a guide is given in figure 5.29. The capacity calculations derive from the spatial considerations. Thus volume of soil to be compacted and manoeuvring space are important factors when choosing compaction plant. Again the best explanation of this is by considering an example. A self-propelled, double-drum smooth-wheeled roller has:

$$\begin{aligned}
\text{roller width} \quad &= 2.03 \text{ m} \\
\text{weight} \quad &= 3000 \text{ kg/m width of roller} \\
\text{speed} \quad &= 2 \text{ km/h}
\end{aligned}$$

From table 5.5 we can see that for the above roller and a well-graded granular material the maximum thickness of compacted soil = 125 mm

$$\text{Minimum number of passes per layer} = \frac{8 \text{ (From table 5.5)}}{2 \text{ (for a double drum)}} = 4$$

$$\begin{aligned}
\text{Area covered} \quad &= (2.03 \text{ (roller width)} - 0.15 \text{ (150 mm overlap)}) \times 2 \times 10^3 \\
&= 3760 \text{ m}^2\text{/h}
\end{aligned}$$

$$\text{Volume of soil handled per hour} = \frac{3760 \times 0.125}{4} = 117.5 \text{ m}^3\text{/h}$$

Volume handled per week = 117.5 × 40 *(40 hour week)*

Volume per week = 4,700 m³/week

## 5.4 SLOPE STABILITY

We have already considered excavation stability to some extent within the chapter on ground water control. We shall now consider slope stability in more detail for both excavations and embankments.

### 5.4.1 Slope Stability of Excavations

The batter of a slope must be designed by a suitably qualified Civil Engineer, but the table shown in table 5.6 is a guide to slope batters in relation to soil type and is intended to give the reader a feel for the correct slope. The slopes are described as 'one in two' or '1:2'. This means that the slope gains in height by one metre for every two metres travelled horizontally. There are other means of defining a slope, as shown in figure 5.34, and we must be clear about which we are using.

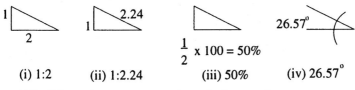

(i) 1:2     (ii) 1:2.24     (iii) 50%     (iv) 26.57°

(All of these slopes are approximately the same)

Figure 5.34

All of the methods shown in figure 5.34 can be used to define a slope, although the percentage system is usually used when defining road gradients. In figure 5.35 a material of well jointed or bedded rock is included to draw attention to the fact that a slope constructed in such a material would have to follow the joints in the rock. For example, a trench cut in slate with a 60° dip would need to have the sides of the trench cut at 60° if it is to be safe.

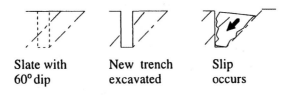

Slate with          New trench          Slip
60° dip             excavated           occurs

Figure 5.35

Finally, in the table 5.6 the reader will notice, perhaps with some surprise, that the temporary slope for clay is steeper than the permanent slope. This is because clay is stronger in the short term, due to cohesion. Initially, clay is held together by cohesion and so can stand at steep gradients. Slowly, however, pore water pressures equalise, the effect of cohesion becomes negligible and intergranular friction dictates the strength of the material, as though it is granular. The clay must then take up a much shallower slope.

Table 5.6

| Material | Slope Batters * (vertical: horizontal) | |
| | Permanent**** | Temporary |
| --- | --- | --- |
| Solid rock (Massive ) | 1.5:1 to vertical | 1.5:1 to vertical |
| Well jointed/ bedded rock | 1:1 to 2:1** | 1:1 to 2:1** |
| Gravel | 1:2 to 1:3 | 1:2 to 1:3 |
| Sand | 1:2 to 1:3 | 1:2 to 1:3 |
| Clay | 1:6*** to 1:5 | 1:2 to 2:1 |

*Note that this table is a **GUIDE ONLY** and should hot be taken as a final design.
**Note this slope will be dictated by the angle of the joints in the rock.
***Note that the clay slope is greater in the short term than it is for the long term as shear resistance is greater in the short term due to cohesion.
****Note that all permanent slopes must be designed for weathering.

For all slope conditions weathering will have a significant weakening effect in the strength of the soil. Obviously, weathering of the slope in the long term will have the greatest effect and should also be taken account of in the design, see section 5.4.4.

### 5.4.2 Stability Calculations

Excavations can be temporary or permanent, but in either case to be safe they must remain stable throughout their design life. Stability of cuttings can be achieved by excavating to a stable slope. The stability of a slope is determined largely by the density and shear strength of the soil of which the slope is constructed. Ground water conditions, pore water pressure and discontinuities within the soil structure also affect the stability, but these are beyond the scope of this book and for more detailed information see Barnes[1]. This text will consider slope stability due to soil density and internal shear strength only.

The stability of an embankment is determined by the relative magnitude of two opposing forces which are the weight of the soil (destabilising force) and the internal shear strength (stabilising force), see figure 5.36. The plane of failure is usually a circle, the position and radius of which can only be found by trial and error.

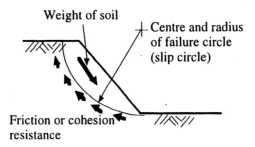

Figure 5.36

Consider a slope of 1:1 and a height of five metres in a clay soil of cohesion value 20 kN/mm² and a density of 18 kN/m³. Find the factor of safety for stability in the temporary condition. Here we have assumed a slip circle radius of six metres, located directly over the toe of the slope. If we look at a one metre run of slope, then the effect of two opposing forces is calculated by taking moments about the centre of the slip circle, (see figure 5.37).

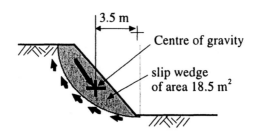

Figure 5.37

Destabilising (or Active) force = 18 (*density*) × 18.5 (*the area within the slip circle*) × 3.5 (*horizontal distance from the centre of gravity of the slip wedge to the centre of rotation*) = 1165.5 kNm

Stabilising force = 20 (*cohesion (ignoring the internal angle of friction)*) × 8.5 (*length of slip circle*) × 6 (*slip circle radius*) = 1020 kNm

Factor of safety = 1020/1165.5 = 0.875

Here we see that the destabilising forces are greater than the stabilising forces which indicates that the slope is unstable. We need to achieve a factor of safety of at least 1.5 if we are to consider the slope to be stable.

We have only considered one slip circle of radius six metres with the centre positioned above the toe. This may not be the radius or position that will give the lowest factor of safety and we must spend time considering different radii circles and different positions of the centre of the circle until we are satisfied that we have the lowest factor of safety. In this case the active forces are so much greater than the stabilising forces that we can conclude that a shallower slope is required for this soil. We have only considered a clay soil in the temporary condition because we have considered cohesion only and have not taken account of the internal angle of friction. If we were to consider a soil of internal friction forces only then we would have to use a different method of calculation called the 'method of slices' (see Barnes[1]).

### 5.4.3 Slope Stability of Embankments

An embankment must be designed by a suitably qualified Civil Engineer, however, table 5.7 is a guide to slope batters in relation to soil type to give the reader a feel for the correct slope.

Table 5.7

| Material | Slope Batters* (vertical: horizontal) |
|---|---|
| Hard rock fill | 1:2 to 1:3 |
| Weak rock fill | 1:2 to 1:3 |
| Gravel sand clay | 1:2 to 1:3 |
| Sand | 1:2 to 1:3 |
| Clay | 1:5 to 1:6 |

*\* This table is a GUIDE ONLY and should not be taken as a final design.*

The type of fill material used to construct an embankment will largely depend upon the material most economically available at the time. Whatever material is used it must fulfil a number of criteria.

- Stability of embankment (design)
- Bearing capacity of the embankment (CBR)
- Settlement of the embankment (due to the weight of fill and the overburden)
- Trafficability of the embankment (constructability)

First we must establish if the material is strong enough, both the short and long term, to construct the slopes required and this of course we do by design. The method of design would be similar to that described previously, but would have the addition of a load at the top of the slope called surcharge, due to the presence of the road and traffic. Once the embankment is formed the material must be strong enough to support a road pavement. This is measured with the California Bearing Ratio (CBR) test, see Chapter 10.

The material must also be assessed for settlement under its own weight and the weight of construction activities. For this reason it is common to overfill an embankment above the chosen formation level so that the material can settle. Another factor to take into account is the settlement of the soil upon which the embankment is constructed. Obviously, this soil is now subject to a considerable extra load called an overburden due to the embankment and is bound to settle under this weight. Finally, the material chosen for the construction of an embankment must be robust enough to be trafficked by construction plant and not be too susceptible to the weather.

### 5.4.4 Weathering

A very important aspect of slope design is the consideration of weathering. Stabilisation of embankments against weathering is achieved by:

- Design
- Drainage
- Vegetation

When the slope is designed for long term stability, it must also be designed against weathering. Surface water can seriously erode a slope and must be controlled by drainage. V-ditches are used at the top of the slope and French drains at the bottom. Sometimes it is necessary to provide herring bone drainage or French drains vertically down the slope to aid drainage of clay. Areas of soft clay must be replaced with more stable material. Stabilisation of the surface is achieved by planting vegetation; this reduces erosion and binds the surface which will be most susceptible to negative pore water pressure problems.

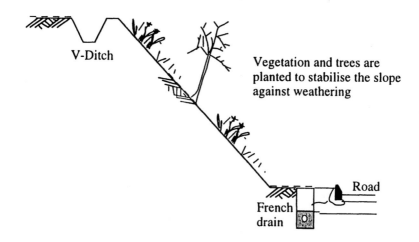

Figure 5.38

## 5.5 REFERENCES

1.  G. Barnes, *Soil Mechanics,* 1995 Macmillan Press, ISBN 0-333-59654-4
2.  British Standards Institute. *BS 1377 : Part 1 to 9: 1990: British Standard*

*Method of test for Soils for Civil Engineering Purposes.*
3.  Department of Transport, *Manual of Contract Documents for Highway Works. Vol 1 and 2,* 1993 published by HMSO.

## 5.6 PROBLEMS

1.  Table 5.8 shows the volumes of cut (positive) and fill (negative) for each 100m length along a proposed road 1.2 km long.
    a)  Plot on to graph paper at a suitable scale the mass profile and mass haul diagrams.
    b)  From the mass haul diagram, determine the surplus which is to be removed from the site at chainage zero.
    c)  Without calculation suggest suitable plant teams for the job, assuming that the work will be carried out at least partly during the winter months.

2.  Figure 5.39 shows the volume of cut (+ve) and fill (−ve) for each 100 m length required for the construction of a new road.
    a)  Plot on to graph paper the mass profile and the mass haul diagram.
    b)  Indicate the surplus obtained at chainage 600, the direction of haul and the balance line.

3.  Table 5.9 below shows the volume of cut (+ve) and fill (−ve) for each 100 m length required for the construction of a new road.
    a)  Plot the mass haul diagram on the graph paper provided.
    b)  List the information that can be gained from the mass haul diagram.
    c)  Explain how the 'free haul' can be used as a tool in the design and construction of roads.

4.  A road cutting is to be constructed through a hill site in Sussex. The cutting is to be about 10 m deep, in chalk with batters in the order of 1:2 slope.
    a)  Table 5.10 gives the quantities of clay soil which must be moved in summer to construct the above road. Any surplus will be taken to the tip 2 km away by road.
    b)  Draw up the 'mass haul diagram' on graph paper. Mark on the diagram the 'free haul', the 'line of balance', 'direction of haul', the surplus and where it will be taken from.

Table 5.8

| Chainage (m) | Volume ($\times 10^3$ m³) |
|---|---|
| 0 to 100 | +4.2 |
| 100 to 200 | +5.6 |
| 200 to 300 | +3.2 |
| 300 to 400 | −1.8 |
| 400 to 500 | −4 |
| 500 to 600 | −9.2 |
| 600 to 700 | −9.4 |
| 700 to 800 | −4.8 |
| 800 to 900 | +2.2 |
| 900 to 1000 | +7.8 |
| 1000 to 1100 | +7.0 |
| 1100 to 1200 | +5.6 |

Table 5.9

| Chainage (m) | 0 | 100 | 200 | 300 | 400 | 500 | 600 |
|---|---|---|---|---|---|---|---|
| Volume (m³) | 2050 | 4825 | 3300 | zero | −3800 | −3275 | |

Table 5.10

| Chainage (m) | 0 | 100 | 200 | 300 | 400 | 500 | 600 | 700 | 800 | 900 | 1000 |
|---|---|---|---|---|---|---|---|---|---|---|---|
| $\times 10^2$ (m³) | 200 | 350 | 125 | −20 | −100 | −75 | −25 | 150 | 125 | 20 | |

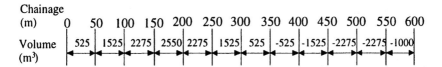

| Chainage (m) | 0 | 50 | 100 | 150 | 200 | 250 | 300 | 350 | 400 | 450 | 500 | 550 | 600 |
|---|---|---|---|---|---|---|---|---|---|---|---|---|---|
| Volume (m³) | | 525 | 1525 | 2275 | 2550 | 2275 | 1525 | 525 | -525 | -1525 | -2275 | -2275 | -1000 |

Figure 5.39

# 6 Foundation Construction and Design

Foundation construction is vital to the success of the project and for this reason much effort is channelled into the foundation design and construction. An important message which must be given to the Client from an early stage is that money spent on the foundations is money well spent. It must be considered as an investment in the structural integrity of the building which will pay dividends well into the future. The consequences of a foundation failure can be huge and even threaten personal safety; for this reason a Chartered Civil Engineer must be employed to gain sufficient information about the ground conditions, design the foundations and oversee the construction.

This chapter looks at the construction process of simple foundations and gives examples of foundation design. Initially, we need to understand the physical behaviour of soils in the context of foundation performance and so we will look at the physical properties of soil and how allowable ground bearing pressures are found. Then the design of pads and piles are considered in some detail. Towards the end of this chapter we shall give an overview of the foundation design process bringing in the soil investigation and how the soil conditions are matched to the most appropriate type of foundation. The chapter concludes with a brief look at ground improvement techniques and underpinning.

The central theme throughout this chapter is the foundations for building structures, but the principles can be applied to any other type of structure which needs a foundation.

## 6.1 TYPES OF FOUNDATION

The purpose of a foundation is to spread the loads from a structure evenly on to the ground in such a way that the ground is able to support the loads applied. There are a number of different types of foundation designed to do this job for different loading and ground conditions. Broadly speaking these are:

- Strip or trench-fill footings used for light line-loading, usually domestic. 15-90 kN/m run.

- Pad foundations used for medium point loads or light loads on poor to good ground. 50-600 kN per pad.

- Raft foundations used for light to medium loading on poor ground.

- Piled foundations used for medium to heavy point loading 600-3000 kN per pile where good ground is only found at depth.

### 6.1.1 Strip Footings

The strip footing is used for low-rise brick built construction. The principle behind its operation is that load concentrations in the brick structure are spread out within the walls so that by the time it reaches the foundation level all that has to be done is to spread the linear load from the wall out sideways. The widths of such footings are rarely designed but are dictated in standard details within the Building Regulations. Of course they can be designed but the loading conditions are standard and provided that there are no ground complications they are often a standard width. Strip footings are usually founded at a depth of about 1.2 m as shown in figure 6.1.

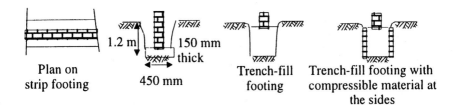

Plan on strip footing    1.2 m    150 mm thick    450 mm    Trench-fill footing    Trench-fill footing with compressible material at the sides

Figure 6.1 Strip footings

The advantage of this footing is that it is cheap, but it is often slow to construct brick or block walls in a trench and so another type of strip footing is used, the 'trench-fill footing'. This type of footing simply fills up the trench with concrete to just below ground level and is ready to receive brick and block construction. See figure 6.1.

The Engineer must be wary in this situation because the trench-fill footing will be vulnerable to sideways pressures from expanding or shrinking soil especially if tree rootlets are present within the clay. The theory is that the soil under the house may dry out and shrink allowing soil outside the house to move the footing inwards. This problem is easily overcome by inserting a compressible board (75 mm thick) in the sides of the trench thus harmlessly taking up any expansion in the soil without allowing pressure to build up on the footing. Alternatively, if the soil under the building is desiccated by tree roots the soil may take up water when the structure is built and an outward pressure be exerted on the footing. This again can be overcome by using a compressible board on the inside of the trench. Trench-fill footings can be provided with steel reinforcement to span over localised areas of ground weakness or any underground services.

### 6.1.2 Pad Foundations

Pad foundations are used to spread concentrated loads such as that from a column. As with strip footings they are often described as shallow footings because they are commonly founded between one and 1.5 m depth. See figure 6.2. Pads are simple and cheap to construct provided that they are founded above the water table, but do not perform well in poor ground. Pads are dependent on the integrity of the soil on which they sit for stability. If localised poor ground conditions are encountered then a different degree of settlement may be encountered on each base. This is called differential settlement and can endanger the stability of a structure. The problem can be overcome by either linking the pads with ground beams or only using pads in reliable, known ground conditions. Isolated pads must be loaded in the centre otherwise an uneven contact pressure is set up which in itself can cause differential settlement. This can be designed for but it is common to overcome the problem by combining bases or using a raft construction (see section 6.1.3).

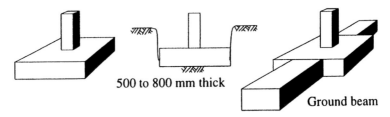

500 to 800 mm thick

Ground beam

Figure 6.2 Typical pad arrangement

Pads are best linked together with some form of ground beam to provide stability against such problems and the ground beam is often required anyway to support the walls of the structure. See figure 6.2. Under straightforward conditions pads are easy to design. Once the load moves off the centre of the base, however, problems are encountered and this is why pads should always be designed by an Engineer.

### 6.1.3 Raft Foundations

Raft foundations are expensive to construct because they use a lot of reinforced concrete which has to be designed and constructed often using shuttering. It is therefore only under unusually poor ground conditions that they are used where all other remedial techniques are not feasible. The principle of operation is similar to that of a raft 'floating' on the soil. It is designed to be very stiff so that it can eliminate differential settlement whilst keeping the structure it supports intact. To increase the stiffness of such a foundation it is common to construct voids or cells within the raft as shown in figure 6.3. This allows the construction to be light and not over-stress the ground. Such cavities must of course be ventilated to prevent the build up of ground gases such as radon and methane.

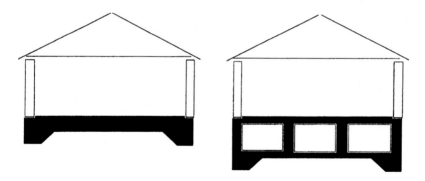

Figure 6.3 Raft Construction

## 6.1.4 Piles

Piled foundations consist of concrete or steel columns placed in the ground and are used when the upper layers of soil are too weak or uncertain to support pad constructions or when there are heavy loads to support and/or minimum settlement is required.

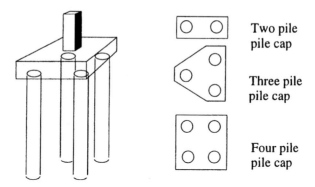

Two pile
pile cap

Three pile
pile cap

Four pile
pile cap

Figure 6.4

As shown in figure 6.4 piles are often installed in groups of two, three or four and are joined together with a pile cap at the top. Pile caps are in turn joined together with ground beams. The primary function of the pile cap is to spread the point load from the bottom of the column of the structure to the tops of the piles and acts as a deep beam spanning between supporting piles. Ground beams then span between adjacent pile caps. The ground beams have basically two functions, the first of which is to support the perimeter walls and edges of floor slabs and the second is to provide extra stability to the pile caps to prevent lateral displacement under load.

## 6.2  SOIL PROPERTIES

Before we can consider foundation design we must understand a little about the physical properties of different soil types. This section will give an overview of different soil types and show how allowable ground pressures are derived. Successful foundation design relies upon a comprehensive knowledge of the physical properties of the ground upon which the foundation will stand. Foundations cannot be designed correctly without a full soil investigation for the proposed site and this is looked at in more detail later. A soil investigation will yield bore hole logs such as those shown in figure 6.32 which classify the types of soil and their location giving rise to a soil profile. There are three broad soil classifications:

- Fine grained
- Coarse grained
- Rock

### 6.2.1  Fine Grained Soils

Fine grained soils consist of silts and clay. Clay has a particle size of less than 0.002 mm which can only be seen with a microscope. Silt particle size varies between 0.002-0.06 mm and can just be seen with the naked eye. The volume and strength of fine grained soils depends on water content. The more water in a clay the greater the volume and the lower the strength; with less water content the clay contracts and the strength increases, this is a fundamental property of the material. The structure of clay consists of very fine plate-like particles rather like microscopic corn-flakes suspended in water and only touching occasionally. The water finds it difficult to move through this structure and so the particles are held in suspension. This is the force that sticks the particles of clay together, called cohesion and, until there is any significant movement of water, called equalisation of pore water pressure, cohesion is the predominant shear strength of the material. Equalisation of pore water pressure will occur over a period of time and as it does the predominant factor determining shear strength will become friction between particles that touch.

Clay can then be summarised as having two types of strength, short term strength and long term strength. In the short term cohesion is the predominant strength and it will remain the same irrespective of the stress applied either due to the weight of soil above called 'overburden' or due to foundation load. In the long term, however, internal friction (the friction between particles that touch) will become the predominant strength. This strength does vary depending upon applied stress due to overburden or foundation load. The more applied stress the greater is the internal strength.

The strength of a soil is called its 'shear strength' and it is this property which we must assess so that we can design a foundation. Under normal conditions the shear strength of the soil is a combination of cohesion and internal friction between particles and is measured by the tri-axial test; the 'undrained' test measures short term strength whilst the 'drained' test measures long term strength, see  G. Barnes[7] The tri-axial test gives the cohesion value or 'C' value (given in $kN/m^2$) and the internal angle of friction (given in degrees). It is common for the C value alone to give a indication of the strength of a soil. Table 6.1 is designed to give the reader a 'feel' for C values in relation to soil description. These values are indicative only, the tri-axial test must be carried out on samples of clay to determine the correct C value. From the C value we can determine the Allowable Bearing Pressure (ABP) and it is this figure that is used in the design of foundations. For more information about soil and how it affects foundation design see BS 8004[4].

Table 6.1 Allowable Bearing Pressures for Clay

| Material Description | Field indications | Undrained Shear Strength $(kN/m^2)$ | Allowable Bearing Pressure $(kN/m^2)$ |
|---|---|---|---|
| Very stiff or Hard | Brittle or very tough | over 150 | 300 |
| Stiff | Cannot be moulded in the fingers | 100-150 | 200-300 |
| Firm to Stiff | | 75-100 | 150-200 |
| Firm | Can be moulded in the fingers by strong pressure | 50-75 | 100-150 |
| Soft to Firm | | 40-50 | 80-100 |
| Soft | Easily moulded in the fingers | 20-40 | 40-80 |
| Very soft | Exudes between the fingers when squeezed in the fist | under 20 | under 40 |

*Note 1 the above ABP's must be divided by two if ground water is present within a depth B of the underside of the proposed foundation.*
*Note 2 All Clays are susceptible to long term settlement which must be taken into consideration in the design.*

Any assessment of ABP must take into account consolidation. Consolidation is a reduction in volume due to the expulsion of water from within the structure of the soil which in turn is due to the applied load. We have already said that water finds it difficult to move through the clay and because of this consolidation occurs over a long period of time, between 6 months to 2 years. The ABPs shown in Table 6.1 are calculated so that consolidation will not exceed 25 mm in the medium term. The values shown for ABPs must be halved if the water table is within 'B' (breath of the base) of the underside of the foundation. For example if a pad base was to be three metres long and two metres breadth then B equals two metres. If the ground water level is one metre below the underside of the pad then the ABP would have to be half of that shown in table 6.1. It would in fact be the same value as the undrained shear strength.

General settlement of a structure will always occur. If it is of a consistently small magnitude (up to 25 mm) then commonly settlement is not a problem. Problems do occur, however, when settlement is not consistent across a site. For example if settlement of 25 mm occurs at the ends of a building and not in the centre, then cracks will begin to appear in the structure. This is called 'differential settlement' and when severe it can endanger the stability of the structure. Such differential settlement can be caused by a varying thickness of clay across the site or by layers of varying strengths of clay. Consequently we can see the importance of not only establishing what the soil profile is but also whether it is consistent across the site. This underlines the importance of carrying out a comprehensive soil investigation before design work can begin.

As an example consider a soil consisting of stiff clay except for a layer of soft clay one metre thick at a depth of three metres. Obviously settlement can be expected of both clays and the soft clay would be expected to consolidate the most. Calculations must then be carried out to estimate the amount of settlement that can be expected for a certain contact pressure. Calculations for this are beyond the scope of this book and the reader is referred to G. Barnes[7]. Also we must examine the shear stresses that will be applied to the soft clay to make sure that this will not fail. An estimate of the shear stress that is applied to the lower layers of soil is given in figure 6.8.

### 6.2.2 Coarse Grained Soils

These soils have negligible cohesion and high permeability. Particle sizes are:

| | |
|---|---|
| Sand particle size | 0.06-2 mm |
| Gravel particle size | 2-60 mm |
| Cobbles particle size | 60-200 mm |

The shear strength is derived only from the friction between particles of soil and is not affected by cohesion. Rapid water flow may however cause fines to be washed out of the soil structure, which reduces density and produces voids which can collapse thus reducing ABP. The strength of the soil can be measured by the shear box test in the laboratory or by the standard penetration test in the field, see BS 1377[5]. The SPT test is where a 35 mm diameter sampling tube is driven 300 mm into the ground with a standard weight hammer. The number of blows required to do this is called the N value and is directly related to the ABP. Table 6.2 gives a range of SPT numbers and shows how they relate to soil type and ABP.

Table 6.2 Allowable Bearing Pressures for Gravel

| Gravel Type | Allowable Bearing Pressure kN/m$^2$ | | | | N value |
|---|---|---|---|---|---|
| | Foundation width m | | | | |
| | 1 | 2 | 3 | 4 | |
| Well graded | 640 | 580 | 540 | 520 | 50 |
| and compact | 510 | 470 | 430 | 410 | 40 |
| | 370 | 350 | 320 | 300 | 30 |
| Medium dense | 260 | 240 | 210 | 190 | 20 |
| | 110 | 100 | 90 | 80 | 10 |
| Poorly graded and loose | 50 | 40 | 35 | 30 | 5 |

These values are not halved if ground water is present within B of the underside of the foundations since the affect of ground water would have already been taken into account by virtue of the fact that the test is carried out in situ. In non-cohesive soils the rate of compression will match the rate of building construction. About 85 per cent of compression will take place before the structure is built so there is no need to consider compression in this circumstance. Magnitude of settlement is limited by design to 25 mm.

Table 6.3 Allowable Bearing Pressures for Rock

| Type of Rock | Allowable Bearing Pressure in kN/m$^2$ |
|---|---|
| Hard igneous and Gneissic rocks in sound condition | 10,000 |
| Hard limestone and sandstone | 4,000 |
| Schist and slates | 3,000 |
| Hard shales, hard mudstones and soft sandstones | 2,000 |
| Soft shales and soft mudstones | 600-1,000 |
| Hard sound chalk (undisturbed), Soft limestone | 600 |
| Thinly bedded limestones, sandstones and shales | To be assessed on inspection |
| Heavily shattered rock | ditto |

*Note These figures assume that the foundations are carried down to unweathered rock.*

### 6.2.3 Rock

The strength of rock is rarely affected by the pressure of water and relies for its shear strength on intermolecular bonding. The strength of rock can be established by crushing cube or cylinder samples as in the concrete cube test. We must, however, exercise care when considering weaker rocks such as slate, limestone or sandstone. Slate is naturally deposited in layers and the layers may have been distorted so that they have an angle of dip. Applied stresses will naturally follow such layers causing a distortion of stress distribution within the soil. (See drainage excavation 9.3) Limestone and sandstone, on the other hand, are prone to cavitation caused by the movement of ground water. Such cavities must be investigated and if necessary filled with grout. It may be an advantage for the construction of piles in rock to use an in situ concrete construction so that any cavities found are automatically filled with concrete. Table 6.3 gives an indication of allowable bearing pressure that can be used for different types of rock. These figures assume that the foundations are carried down to unweathered rock.

### 6.2.4 Contact Pressure

This is the term used for pressure delivered to the soil by the foundation. This is usually considered as uniform for the purposes of design but is in fact distributed differently in cohesive to non-cohesive soils. This can best be explained by first considering a footing which is flexible but applies a uniform load over its entire area. Such a foundation would take up a convex shape on sand or any cohesion-less soil and a concave shape on clay see figure 6.5.

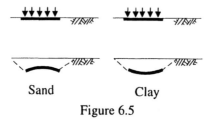

Sand                    Clay

Figure 6.5

The reason for this behaviour in sand is that there is no overburden pressure at the edges of the footing to give the sand the shear strength it needs to support the loads, whereas the sand beneath the centre of the footing is confined and rapidly gains strength as the load is applied. The result is that there is settlement movement at the edges of the footing with little movement in the centre and the convex shape is formed. In clay the story is different because strength is independent of overburden pressure. Compression of the soil at the centre of the footing can be one and a half times that at the edges and thus the concave shape is formed. In reality we consider foundations as being rigid and they are designed not to deform to the shape of the soil. This means that the contact pressure beneath the foundation will vary depending upon the type of soil on which it sits. In clay a rigid footing loaded with a uniform load P will have a contact pressure of about 0.5P at the centre and in theory increase to the yield pressure at the edges. This is shown in figure 6.6. In sand the maximum pressure is found at the centre of the footing and reduces in parabolic form to zero at the edges. This shape becomes less apparent as the footing gets wider and becomes more uniform.

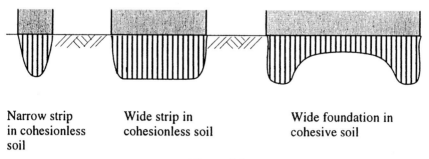

Narrow strip        Wide strip in              Wide foundation in
in cohesionless     cohesionless soil          cohesive soil
soil

Figure 6.6

In reality soils are usually a mixture of cohesive and non-cohesive material and foundations are semi-rigid. This means that the designer is more than justified to assume that contact pressure is distributed evenly across the footing no matter what type of soil.

### 6.2.5 Stress Distribution

The distribution of stress within the soil, found by Jürgenson[8] in 1934, reduces radially with depth and distance from the centre as shown in figure 6.7 and 6.8. It should be noted that the stresses are in addition to those induced by the weight of the soil itself. Knowledge of these stress distributions help the Engineer to assess the effect of loads on lower layers of soil, especially if they happen to be weaker than the upper layers. For example, if a weak clay existed 1.5 m below a square foundation three metres wide bearing on a clay of C value 60 kN/m², we can see from figure 6.7 that approximately 0.3 of the contact pressure will be applied to the weaker layer as shear stress. If a load of 100 kN/m² is applied at the surface, then 30 kN/m² is applied to the weaker soil. The shear strength of the weaker soil may not be sufficient to support this in addition to the stresses induced by the weight of the soil above, so we may have to reduce the contact pressure by increasing the area of the footing. Most foundations are square or rectangular pads, but figure 6.7 is for a circular base and figure 6.8 is for a strip footing. For design purposes the stress distribution for the square base can be approximated to that of a circular base, and a rectangular base three times longer than its width, can be approximated to a strip footing. Plastic failure of soil will occur when the ultimate bearing pressure for soil is reached. See figure 6.9. The failure mechanism is usually shear and is likely to occur when the bearing pressure is approximately six times the shear strength of the soil.

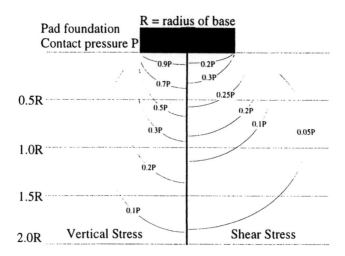

Figure 6.7 Typical pressure bulbs for a circular footing

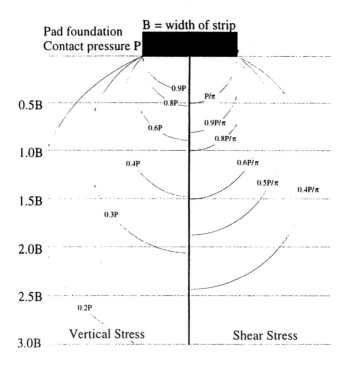

Figure 6.8 Typical pressure bulbs for a strip footing

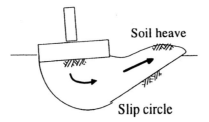

Figure 6.9 Plastic failure of a pad foundation

## 6.3 PAD DESIGN AND CONSTRUCTION

Pad foundations are designed to support point loads from the structure and spread them to the soil. The Engineer must design the size of pad so that the applied point load is spread evenly onto the ground and this section looks in detail at pad design. All foundations exert a pressure onto the ground and this is known as the contact pressure. The contact pressure must not be allowed to exceed the allowable ground bearing pressure (ABP).

The point load may be applied at any position on the pad, when the load is central the pad is said to be axially loaded. Under axial loading conditions the pad design is straightforward. The load is divided by the area of the base to calculate the contact pressure. The contact pressure must then be checked to confirm that it is not greater than the allowable ground bearing pressure.

*Example*

Consider a 200 kN load to be supported by a pad footing sitting on a clay which has a cohesion value of 75 kN/m². Find the size of pad required.

Assuming that there is no ground water present within a depth equal to the width of the proposed pad an ABP of 2 × 75 = 150 kN/m² can be used in the design. (This ABP assumes a maximum settlement of 25 mm.)

$$\text{Area of pad required} = 200/150 = 1.33 \text{ m}^2$$

Pad size = $\sqrt{1.33}$ = 1.15 m² (allowing maximum settlement of 25 mm)

On the drawing we would call up a **1.2 m square pad.**

The contact pressure can be varied by changing the size of the base and the designer must aim to achieve a standard contact pressure for all pads on the same structure. This will ensure that approximately the same degree of settlement will occur on each base, thus avoiding differential settlement problems. This approach will give rise to many different size bases but it is usual to standardise pads to two or three sizes on a job.

### 6.3.1 Eccentric Loading

It is very often the case that a column for a structure cannot sit directly in the middle of a pad footing due to some restriction to one side from say a site boundary or existing foundations to an adjacent building. This causes a triangular distribution of contact pressures which at the highest point can exceed the allowable ground bearing pressure. See figure 6.10. Such a loading configuration must be considered carefully.

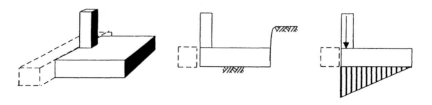

Figure 6.10

The danger of an uneven distribution of contact pressures is that the soil experiences varying degrees of compression which may cause differential settlement problems. Such loading variations can be designed for; consider figure 6.11. First we must check that $S_2$ is always positive: if it is, then lifting of one side of the base will not occur. We do this by checking that the line of action of the applied loads is within L/6 of the centre of the base. This will ensure stability. Then we calculate the contact pressures $S_1$ and $S_2$. We must ensure that $S_1$ does not exceed the ABP and that the variation in pressure is not too great. This is done by limiting the variation to say 40 kN/m² for an ABP of 120 kN/m². If the above limits cannot be achieved then we must consider the use of balanced or cantilever foundations.

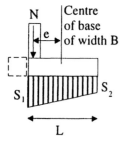

when $\dfrac{Ne}{N} < \dfrac{L}{6}$      $S_1 = \dfrac{N}{BL} + \dfrac{Ne}{z}$      $S_2 = \dfrac{N}{BL} - \dfrac{Ne}{z}$

Where   N = Axial load from column
        e = Eccentricity of load from centre of base
        L = Length of base
        B = Width of base
        z = Section Modulus = $\dfrac{BL^2}{6}$

Figure 6.11

*Example*

Consider a 200 kN load to be supported by a pad footing sitting on clay which has a cohesion value of 50 kN/m². The load is eccentric from the centre of the base by 150 mm in one direction. Find the size of pad required.

Assuming that there is no ground water present within a depth equal to the width of the proposed pad an ABP of 2 × 50 = 100 kN/m² can be used in the design. (This ABP assumes a maximum settlement of 25 mm.) Try a two metre square pad.

$$\frac{Ne}{N} = \frac{200 \times 0.15}{200} = 0.15$$

$$\frac{L}{6} = \frac{2}{6} = 0.33$$

Thus $\frac{Ne}{N} < \frac{L}{6}$ which means that there will be no uplift and is acceptable.

$$S_1 = \frac{N}{BL} + \frac{Ne}{z} \qquad S_1 = \frac{200}{2 \times 2} + \frac{200 \times 0.15}{\dfrac{2 \times 2^2}{6}}$$

$$S_1 = 50 + 22.5 = 72.5 \text{ kN/m}^2$$

$$S_2 = 50 - 22.5 = 27.5 \text{ kN/m}^2$$

This proves that the contact pressures are within allowable bearing limits but the variation is a little too large. Depending on the ground type we must therefore consider using a balanced or cantilever footing. A granular soil, on the other hand, is not subject to long term settlement and such a variation of contact pressures may be acceptable. For simplicity the above calculations have ignored the self-weight of the base which must be taken into account in the design.

## 6.3.2 Combined Balanced and Cantilever Foundations

Off centre loading, called 'eccentric' loading is often dealt with by using a combined pad (sometimes called a 'balanced base') or a cantilever ground beam. The idea behind a balanced base is to extend the pad to the next adjacent column and design its shape such that the line of action of the applied loads corresponds to the centre of gravity of the combined footing on plan. See figure 6.13. The line of action of the applied loads is found by taking moments about some point that we choose, say A, (see figure 6.12). The contact pressures are then calculated using the same method as above for eccentrically loaded footings. Contact pressures are then compared to allowable bearing pressures and if they are within acceptable limits reinforcement for the footing can be designed.

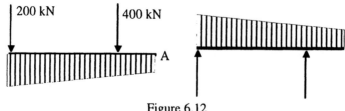

200 kN

400 kN

A

Figure 6.12

The combined pad acts as a wide beam resisting bending and shear forces and spans between the columns. In fact the method of design is very similar to that of a simple beam. If you can imagine the loading arrangement upside down with the columns acting as supports for the beam and the applied load from the contact pressure of the soil, then we can design the base.

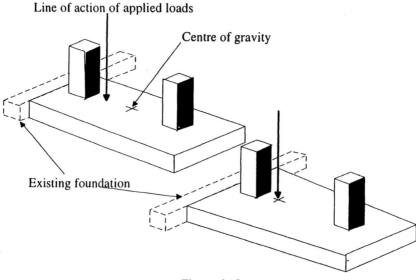

Line of action of applied loads

Centre of gravity

Existing foundation

Figure 6.13

If the contact pressure cannot be balanced within a rectangular base then we can alter the shape of the base to bring the centre of gravity in line with the resultant of the applied loads. See figure 6.13. This will have the effect of balancing the loads so that the contact pressure is uniform. The same equations as shown above are used, the only difference being that the resultant of the applied loads is found by taking moments about one end giving the ideal position for the centre of gravity of the base.

Columns which are too close to existing footings to accommodate balanced footings can be supported by using cantilever beams. The idea behind this design is that the moments set up by the eccentric load are balanced by the downward load of an adjacent column. The beam is designed to support the eccentric load via a cantilever action whilst the pads support the beam. The beam can either be designed to be integral with the pads or rest on top of the pads as shown in figure 6.14.

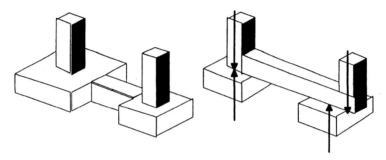

Figure 6.14

*Example 1*

Consider the loading conditions shown in figure 6.15 for a foundation to sit on a clay which has a cohesion value of 100 kN/m² and ground water one metre below the intended underside of the foundation. Find the size of pad required.

Because ground water is present within a depth equal to the width of the proposed pad an ABP of $1 \times 100 = 100$ kN/m² can be used in the design. (This ABP assumes a maximum settlement of 25 mm.)

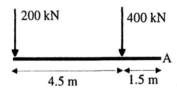

Figure 6.15

First we must find the centre of action of the applied loads and this is done by taking moments about 'A'.

$$200 \times 6 + 400 \times 1.5 = 600 \times \text{'X'}$$

Where X is the distance of the line of action of all the applied loads from A. Solving this equation X = 3 m.

So we can now say that the applied loads are equivalent to a 600 kN load at a distance of three metres from A, see figure 6.16.

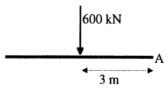

Figure 6.16

This load is at the centre of the base and so the contact pressures are even and assuming that the footing is two metres wide:

$$\text{contact pressure} = \frac{600}{2 \times 6} = 50 \text{ kN/m}^2$$

The contact pressures are less than the allowable ground bearing pressures and so we would simply call up a 2 × 6 m footing on the drawing.

It is not always possible to obtain a uniform contact pressure under a foundation and this next example shows how a varying contact pressure can be designed for.

*Example 2*

Consider the loading conditions shown in figure 6.17 for a foundation to sit on a medium dense, compact but poorly graded gravel, which has an N value of 10. No ground water is present within a depth equal to the width of the proposed pad. Find the size of pad required.

Try a two metre wide footing. From table 6.2 we can see that the allowable ground bearing pressure for a two metre wide footing is 100 kN/m².

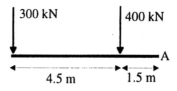

Figure 6.17

First we must find the centre of action of the applied loads and this is done by taking moments about 'A'

$$300 \times 6 + 400 \times 1.5 = 700 \times \text{'X'}$$

where X is the distance of the line of action of the applied loads from A
Solving the above equation gives X = 3.429 m

We can now say that the applied loads are equivalent to a 700 kN load at a distance of 3.429 m from A, see figure 6.18.

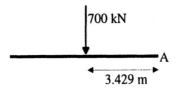

700 kN

A

3.429 m

Figure 6.18

Here the loading is not at the centre of the base, we can see that the eccentricity is 0.429 m and we must check for uplift.

$$\text{check } \frac{L}{6} = \frac{6}{6} = 1.0$$

$$\text{here } \frac{Ne}{N} < \frac{L}{6} \quad \text{i.e. } 0.429 < 1.0$$

This means that there will be no uplift and is therefore acceptable.

$$\text{Contact pressures are:} = \frac{N}{BL} \pm \frac{Ne}{z}$$

$$= \frac{700}{2 \times 6} \pm \frac{700 \times 0.429}{\dfrac{2 \times 6^2}{6}}$$

$$S_1 = 58.3 + 25 = 83.3 \text{ kN/m}^2$$

$$S_2 = 58.3 - 25 = 33.3 \text{ kN/m}^2$$

The final arrangement is as shown in figure 6.19

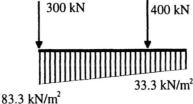

300 kN                400 kN

33.3 kN/m²

83.3 kN/m²

Figure 6.19

These contact pressures are below the allowable ground bearing pressure of 100 kN/m² and so are acceptable. The variation is a little high, but for a gravel is acceptable. Remember that 80 per cent of settlement is likely to occur during the construction period.

Once ground pressure distribution is calculated, pad reinforcement must be designed in accordance with BS 8110[6] and checked for bending, direct shear and punching shear as shown in figure 6.20. As stated previously, this can be accomplished by considering the loading arrangement inverted and so treating the design as a wide beam spanning between point loads. Such a design is beyond the scope of this book.

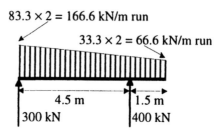

$83.3 \times 2 = 166.6$ kN/m run

$33.3 \times 2 = 66.6$ kN/m run

4.5 m     1.5 m

300 kN     400 kN

Figure 6.20

Note that where new foundations abut existing foundations, the new foundations must be constructed at the same level as the existing foundation so as not to undermine the existing and possibly cause movement.

### 6.3.3 Biaxially Loaded Pads and Raft Design

Biaxially loaded pads are where the load is not in the centre of the pad in the x direction or the y direction. The methods of design are very similar to that shown above. Raft design is similar to the balanced pad design, except that we have to consider loads in two dimensions (x and y). Finding the centre of gravity of applied loading is a little more involved but is still a simple hand-calculation.

*Example*

Consider the loading conditions shown in figure 6.21 on a gravel which has an allowable ground bearing pressure for a two metre wide footing of 240 kN/m².

First we must check that there is no uplift and so check $\dfrac{Ne}{N} < \dfrac{L}{6}$

$$\text{Here } \frac{400 \times 0.15}{400} = 0.15 \text{ in the x direction}$$

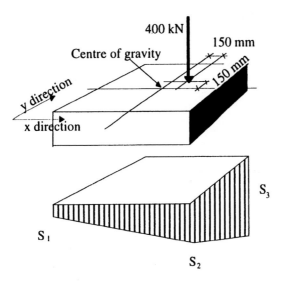

**Figure 6.21**

and $\dfrac{2}{6} = 0.3$ (assuming two metre square base)

so we can confirm that there is no uplift. The calculation is the same for the y direction. The equation that gives the contact pressures is:

$$\text{contact pressures} = \frac{N}{BL} \pm \frac{Ne_x}{z_x} \pm \frac{Ne_y}{z_y}$$

$$= \frac{400}{2 \times 2} \pm \frac{400 \times 0.15}{\dfrac{2 \times 2^2}{6}} \pm \frac{400 \times 0.15}{\dfrac{2 \times 2^2}{6}}$$

$$S_1 = 100 - 45 - 45 = 10 \text{ kN/m}^2$$

$$S_2 = 100 + 45 - 45 = 100 \text{ kN/m}^2$$

$$S_3 = 100 + 45 + 45 = 190 \text{ kN/m}^2$$

Here contact pressure is within the limits set by the allowable ground bearing pressure, but the variations are so large that a balanced or combined footing of some form should be used even for a granular soil.

Raft design is similar to the above except that the loading condition is usually the result of many point and line loads from the supported structure. The Engineer must find the position of the line of action of the resultant load and compare that to the position of the centre of the plan area of the raft. Find the eccentricities in the x and y directions and design as shown above. The position of the resultant force is found by taking moments about a corner position in both the x and y directions. The raft structure itself must then be designed to resist bending and shear forces.

## 6.4 PILE DESIGN AND CONSTRUCTION

Generally piles are used if settlement is a problem caused by soft layers of ground or if the Engineer wishes to limit settlement to a small magnitude. Loads of 300-800 kN/pile are common on piles.

Whatever construction method is used the pile is either predominantly end bearing or friction bearing, see figure 6.22. End bearing piles derive their strength by simply sitting on a hard layer of soil. They are used to transmit load through weak or soft soil to a stronger stratum such as rock lower down and must be capable of carrying the load. Friction piles derive support from friction or adhesion resistance generated by the soil in direct contact with the surfaces of the pile. Most piles in fact gain their strength from a combination of the two.

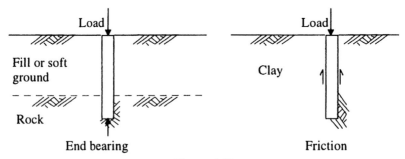

Figure 6.22

For end bearing piles the underlying stratum must be strong enough to resist the inevitable high loads applied. Typical loads of 300-600 kN/pile are common on limestone. It is often the case that the top of the rock layer has undergone some form of weathering in the past and the Engineer must ensure that the rock on which the pile will sit is in a sound condition. For this reason it is common to 'socket' the pile into firm rock by drilling 300 to 600 mm into the rock.

If the end bearing option is not available then the pile can be designed as a friction pile. Friction can only be counted upon once past any weak ground or filled layers. The length of the pile in reliable soil is called the effective length of the pile and it is this effective length which is used for design, see figure 6.23. Although we do not take account of friction resistance from the weaker upper layers of soil they do have some friction and may help the load bearing capacity of the pile. It is more common, however, to protect this section of the pile against developing friction by the use of a plastic sleeve. This is because the soft ground will settle more quickly than the pile and drag down on the pile and actually increase the applied load. This is called 'negative skin friction' and must be avoided at all costs.

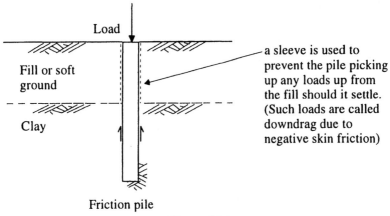

Friction pile

Figure 6.23

### 6.4.1 Pile Design

Piles are generally of two types: displacement or replacement piles, both are shown in figure 6.24. Displacement piles are preformed piles which are hammered into the ground like nails into wood, the soil being displaced sideways to allow room for the pile. This simply compacts the soil allowing it to carry more load. Displacement piles are driven to a 'set', this means that the pile is driven into the soil until a defined number of blows advance the pile into the ground by 25 mm or less. The number of blows depends on the soil but is between 5 to 20 blows.

The advantage of this type of pile is that each is 'tested' by the set so the Engineer can be confident that all piles will take the designed load. Another advantage is that such a pile can be used in contaminated ground without exposing the contaminants to construction workers or the environment. Because the piles are preformed, often under factory conditions, the quality of the pile in the ground can be assured. The disadvantage of preformed piles is that they must be made to set lengths and then transported to site and joined to form the pile. Such joints are weak and if the pile hits an underground obstruction this can break the joints and send the pile off line without any sign on the surface. Another disadvantage is that the installation process is noisy.

Replacement piles are columns of concrete cast in the ground in a predrilled hole. The drilling action removes the soil where the pile is to be and replaces it with concrete. The advantages of this system are that it is relatively quiet and underground obstructions can be dealt with during the drilling. Disadvantages are that all piles cannot be load tested and a separate testing procedure must be instigated. The walls of the drilled hole must also remain stable when

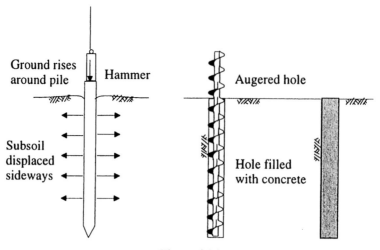

Figure 6.24

unsupported to allow concrete to be placed. The walls of the hole may bulge in during the concrete pour or when the concrete is wet, causing a restriction in the diameter of the pile; this is called necking. If severe, necking can cause the pile to be weak or even fail under load. The solution to this problem is to use a temporary steel casing inserted in the hole during drilling to support the ground whilst concrete in placed and remove it before the concrete gains strength.

*Example*

Consider a 200 kN loading on a pile in clay of average cohesion 80 kN/m². There are two metres of soft peat above the clay.
  Firstly, we recognised that the pile is predominantly a friction bearing pile, bearing in the clay with some form of plastic sleeve over the top two metres to prevent negative skin friction. If we try a pile ten metres long and diameter 350 mm, then the effective length is eight metres. Using Skempton's formula:

$$Q_a = \frac{\alpha PlC + 9A_bC}{\text{Factor of safety}}$$

Where   $Q_a$  = Allowable pile load
         $\alpha$   = Adhesion factor (usually 0.45 for London clay)
         P    = Perimeter of pile shaft
         l    = Effective length of pile shaft for design
         C    = Average cohesion value for pile shaft
         $A_b$ = Area of pile base
         C    = Cohesion value at pile base
         Factor of safety is usually 2.5

$$Qa = \frac{0.45 \times (2 \times \pi \times 0.175) \times 8 \times 80 + 9 \times (\pi \times 0.175^2) \times 80}{2.5}$$

$$Qa = \frac{217 + 69.3}{2.5} = 154.4 \text{ kN}$$

This is not enough and so we must extend the pile to say 12 m long
Effective length is 10 m.

$$Qa = \frac{0.45 \times (2 \times \pi \times 0.175) \times 10 \times 80 + 9 \times (\pi \times 0.175^2) \times 80}{2.5}$$

$$Qa = \frac{395.8 + 69.3}{2.5} = 186 \text{ kN}$$

This is not enough and so we must extend the pile to say 14 m long
Effective length is 12 m.

$$Qa = \frac{0.45 \times (2 \times \pi \times 0.175) \times 12 \times 80 + 9 \times (\pi \times 0.175^2) \times 80}{2.5}$$

$$Qa = \frac{475 + 69.3}{2.5} = 217.7 \text{ kN}$$

This is sufficient to cover the applied load and we would call up 14 m long piles
of 350 mm diameter on the drawings.

This is a simplified design because it is usual for the cohesion value to
increase with depth and we would therefore need to find the average value of
cohesion along the effective length of the pile for the above calculations.

### 6.4.2 Pile Construction

There are many different types of construction available on the market and just a
few common types are discussed here.

*Continuous Flight Auger (CFA)*

This is the most commonly used form of pile. The sequence of construction is
shown in figure 6.25. The installation is carried out by a specialised piling rig
which is brought to site for the job. It can travel on level ground and positions
itself over each pile location. The auger has a central stem through which
concrete can be pumped at high pressure. Step 1 shows the auger being drilled
into the ground. Very little spoil is evident on the surface at this stage as the
rotating action pulls the auger into the ground. Step 2: once the auger has
reached the required depth concrete is pumped down the stem at high pressure
into the bottom of the pile. Step 3: as concrete is pumped into the hole the auger
is slowly raised at a controlled rate whilst maintaining a slow rotation in the drill
direction. The pressure of the concrete assists withdrawal and spoil is raised to
the surface. Care must be taken at this stage not to withdraw the auger too

quickly otherwise a void will form at the end of the auger and the soil may collapse into it. Step 4: when the boring is complete spoil and surplus concrete is cleaned away from the top of the pile, a reinforcing cage is lowered into the still liquid concrete. Step 5: the concrete must be allowed to cure for 7 days to gain full strength when testing can be carried out. The reinforcing cage shown here extends for the full length of the pile, but this is not always the case since generally only the upper portion of the pile is reinforced. Full length reinforcement is used if a pile is expected to be subject to tension or sideways 'lateral' forces.

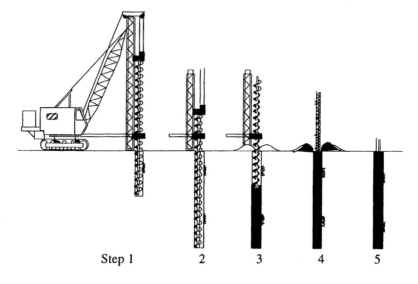

Step 1   2   3   4   5

Figure 6.25 (*Courtesy of Westpile Ltd*)

This is clearly a replacement pile. Its advantages are that it is quiet to install and will not subject the ground to too much disturbance and so can be used close to existing foundations. Continuous flight augered piles are recommended for end bearing piles because they can drill past weathered layers into sound rock and automatically fill any solution cavities encountered. Continuous flight augered piles are not recommended in weak ground or ground which is experiencing large movements of ground water because the concrete must be left undisturbed for a period of time until it has cured. However, in such circumstances a temporary steel casing can be used and withdrawn before the concrete sets. A permanent casing is required for very poor or aggressive ground conditions and in that case a shell pile is required (see later).

*Precast Pile*

Precast piles are cast in factory conditions with high strength concrete and reinforced for their full length. The process of installation is shown in figure

6.26. Step 1: simply position the piling rig in the desired location of a pile and hammer the pile into the ground with either a drop hammer or a diesel driven hammer. Precast piles are brought to site in six metre lengths and joined to each other in the rig by either welding or some form of locking system. Step 2: piles are driven into the ground, end on end, until sufficient set is achieved.

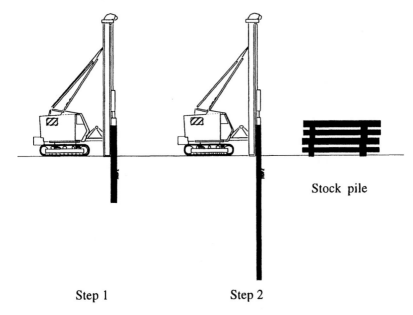

Stock pile

Step 1                    Step 2

Figure 6.26 (*Courtesy of Westpile Ltd*)

Precast piles are displacement piles and are noisy to install. They tend to cause the ground to heave and so cannot be used adjacent to existing foundations. Precast piles are not recommended for fill areas or gravel which may contain large obstructions. This is because they are prone to deflection underground or even cracking if an obstruction is encountered during piling and there is no way of checking for such damage. Also during piling underground obstructions can cause the pile to move off its intended position or move out of plumb, or both. Precast piles are however highly suited to poor ground and can be installed in any ground water condition. Each pile is driven to a set and so can be said to be tested before loading. The construction of the piles is to a high standard of quality using much higher strengths of concrete than can reasonably be used for in situ concrete piles. Precast piles can also be treated or coated to resist most aggressive ground conditions, including ground contaminated by industrial wastes.

*Shell Pile*

Shell piles are a cross between in situ and precast piles, but are essentially a displacement system. The shell system uses an outer concrete or steel shell threaded on to a steel inner core at the end of which is a steel or concrete shoe. The shell and the shoe form the outer shape of the pile and will remain in the ground forming the outside of the pile. The piling force is applied by a conventional hammer and is delivered to the shoe via the steel core. The hammer drives the shell and steel core into the ground and is driven to a set, as for conventional precast concrete piles. Figure 6.27 shows the installation process in progress. Steps 1 and 2: shows the driving of the shell and steel core into the ground. More shell sections and extensions to the steel core can be added to extend the pile until it achieves a set. Step 3: removing the steel core leaves a stable, watertight hollow concrete column. Step 4: reinforcement is lowered into the hole. Step 5: the shell is filled with concrete and compacted forming a solid concrete pile.

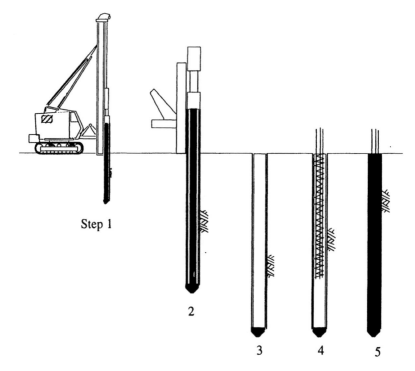

Step 1

Figure 6.27 *(Courtesy of Westpile Ltd)*

This form of pile combines the advantages of the precast system and the in situ system; ease of installation, guaranteed quality of concrete outer skin to resist aggressive ground or ground contamination. The disadvantages of the precast system due to deflection and cracking underground are solved by the steel core. The disadvantages of ground heave and noise however remain, and the system cannot be used where ground water pressures are high.

### 6.4.3 Testing

As has been mentioned before all displacement piles are tested as they are installed by virtue of being driven to a set. This is a form of load testing which does not apply to replacement piles which should be tested either by load testing or integrity testing. A common testing specification for 50 piles would be to load test one pile and carry out integrity testing on the remainder.

*Load Test*

The static load test is carried out by loading the pile to 1.5 × working load for a non-destructive test or to 2.0 × working load for failure. The destructive test is expensive because a special pile has to be installed that will not be used for the structure. Consequently the destructive test is not often used. It is more usual to carry out a non-destructive load test on a working pile. The cost of such a test is high, about £5000 at 1996 prices and so only one such test is carried out on the average job. The process of the test is literally to load up the pile to 1.5 × working load by jacking against heavy weights called 'kentledge' suspended over the pile. Loading is applied in steady increments to design load and released gradually to establish dynamic characteristics and then the pile is loaded to the test limit whilst settlement is measured. Settlement greater than 25 mm under 1.5 × working load is considered a failure. An alternative means of loading the pile is sometimes used by jacking against a beam suspended over the pile and held down by two tension piles. This method of testing is no cheaper than the kentledge test. The load test on its own is also an unreliable test because only one test is carried out in one location and this may not reflect the performance of other piles due to varying ground conditions.

*Integrity Testing*

We cannot load test all piles but the Engineer must satisfy himself that all the piles are in good condition. The integrity test gives an indication of the condition of the pile and is cheap enough to be carried out on all piles, approximately £10 per pile at 1996 prices. Here sound waves are sent down the pile by hitting the top of the pile with a hammer and the time taken for the waves to return is measured by computer. Thus we gain a picture of the length of the pile. Cracks, necking or any obstruction in the concrete of the pile can be readily identified.

### 6.4.4 Groups of Piles

Piles are usually designed with a fixed load in mind so that the piling sub-contractor can use a standard diameter and length of pile. Of course, lengths of piles may vary in accordance with ground conditions. The building designer must look closely at the loads which will be applied to the ground. To do this he must work out all relevant unit loads and calculate their accumulation as the loads are transferred down through the structure, (see section 6.5.1). Once the applied loads are known, they are rationalised to give the pile designer just one or two 'standard' working loads to design for. Load concentrations in the building are then catered for by using two or more 'standard' working load piles. Groups of piles are often used under column and stair or lift positions. Once piles are grouped, this has the effect of reducing the load capacity of each pile in the group by 30 per cent. This is called the bulb effect and must be taken into account in the pile design.

### 6.4.5 Pile Stability

The Engineer must consider stability when determining a pile layout. Stability is an important factor when considering high point loads. The point of application need only be a few centimetres off centre for significant moments or turning forces to be induced which can destabilise a pile. To overcome this problem piles may be installed close to each other forming groups of piles. It is common practice to place piles in groups of two, three or four and connect them with a pile cap. Piles have minimum centres at which they must be placed of approximately two an a half to three times the diameter of the piles used. The pile caps act as deep beams spanning between piles and in this way the misalignment of a heavy point load on a beam is less critical to the stability of the pile group. It is common to use a series of two pile groups connected by ground beams, with the line of the beam at right angles to the centre line of the piles. This arrangement ensures that stability is maintained in two directions. The consideration of stability and the number of piles required to support the applied loads of the structure are the primary factors which determine the layout of piles in a foundation.

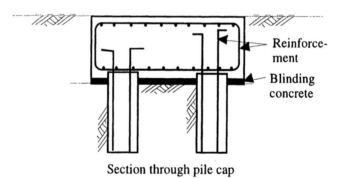

Section through pile cap

Figure 6.28

### 6.4.6 Setting Out Piles

This is usually done by the setting out Engineer. The position of each pile must be accurately measured off grid lines or set out with a theodolite. Piles must be installed within an accuracy of ±75 mm. A steel rod stuck in the ground is usually used to mark the proposed pile position. The site will have been previously excavated down no lower than 600 mm above the correct cut-off level for the pile, so that concrete for the pile can be poured up to the surface. A minimum of 600 mm of pile is always removed down to cut-off level when installation is complete. This is so that any unsound or contaminated concrete can be removed.

### 6.4.7 Pile Caps

Pile caps are constructed of reinforced concrete. The piles are constructed so that some steel reinforcement protrudes from the top, and this is then cast into the pile cap. Typical reinforcement details are shown in figure 6.28.

### 6.4.8 Pile cap Design

Reinforcement can be designed following one of two methods of design, the bending or the truss method. The bending method of design assumes that the pile cap as a simply supported beam spanning between piles, as shown in figure 6.29(i). The truss method of design is shown in figure 6.29(ii) and assumes that the concrete is in direct compression between column and piles and that tension forces between piles are resisted by reinforcement.

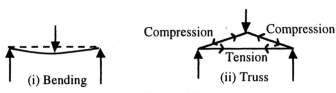

Figure 6.29

For both methods of design, concrete and steel reinforcement must be designed to BS 8110[6]. In the bending method, the shear and local bond stress must be checked for the tension reinforcement. In the truss method the anchorage length and bearing stress on the inside of the tension reinforcement bends must be checked. Both methods of design give similar results.

### 6.4.9 Ground Beams

These are often required to carry intermediate wall and floor loads but are also used to stabilise pile groups. Single and double pile groups require ground beams to give stability under normal loading. See figure 6.30.

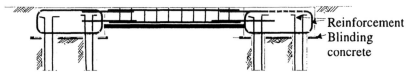

Figure 6.30

## 6.5 FOUNDATION DESIGN

This section looks at the process of design, bringing together a number of different construction techniques and ground considerations. When designing foundations, the Engineer must consider the soil conditions, taking into account the applied loading and cost and match them to the correct foundation type. The Engineer must also take account of any special Client requirements such as minimised settlement and any problems such as ground water, adjacent foundations or weak ground. Ground water control was considered in some detail in Chapter 4, cantilever design earlier in this chapter in section 6.3.2, and ground improvements is considered later in this chapter, in section 6.6.

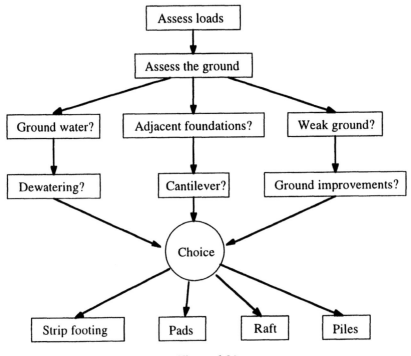

Figure 6.31

No two sites are the same and so even if the same building is to be constructed but in a different location this foundation design process must be carried out on each building. Figure 6.3.1 shows the process of foundation design in the simple form of a flow chart. Designing a foundation consists of calculation of the applied loads, an assessment of the allowable bearing pressure, an estimate of the contact pressures (giving due consideration to settlement effects) and the selection and design of the best foundation type.

### 6.5.1 Assessing Loads

First the Engineer has to assess the applied loads. This is done from an outline scheme drawing supplied by the Architect early in the design process. This drawing gives the Engineer an idea of the building size, layout and materials for construction. The Engineer will be in the process of assessing the best type of structural support for the building and thus, if a framed structure is to be used, will determine beam and column spacing. The weight of the building can then be assessed by considering unit loads for typical wall and floor construction and multiplying these by the area of construction.

Unit loads consist of dead and live loads. Dead loads are the weight of the structure complete with finishes, fixtures and fixed partitions and are supplied in BS 648[1] or direct from manufacturer's literature. Live loads (sometimes called imposed loads) must also be allowed for, and these are found in BS 6399[2]. Live loading takes into account the weight of people and furniture. Wind loading will rarely have an effect on the foundations. Only if the structure is tall and thin will it be significant and these loads can be found from CP3; Chapter V[3].

Once the unit loads have been established, we can estimate the loads on the structure. To do this we must calculate load accumulation as it is transferred down through the structure; this is known as 'chasing down the loads'. Consider the following example.

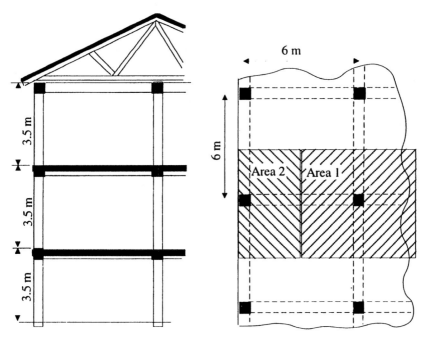

Figure 6.32

Figure 6.32 shows a three-storey structure with a 30° pitched roof of tile and truss construction. The Engineer has decided to use a framed structure of in situ concrete. The cladding is to be brick and block construction; the floors are in situ concrete 150 mm thick on a square grid of drop beams 150 mm below the underside of the slab and 300 mm wide. The beams span between columns 300 mm square at six metres centres. There is a suspended ceiling on each floor but no floor screed.

*Unit Loads*

|        |                                                      | Dead load (kN/m$^2$) | Live load (kN/m$^2$) |
|--------|------------------------------------------------------|---------------------|---------------------|
| Roof   | Interlocking concrete tiles (BS 648[1])              | 0.56                | -                   |
|        | Battens 0.02, insulation 0.04, felt 0.02.            | 0.08                | -                   |
|        | 30° trusses at 600 mm centres                        | 0.14                | -                   |
|        | Suspended ceiling                                    | 0.18                | -                   |
|        | Loft live load (BS 6399[2])                          |                     | 0.25                |
|        | Snow load on roof (live load) (BS 6399[2])           |                     | 0.75                |
|        |                                                      | 0.96                | 1.00                |
| Floors | Concrete floor(BS 648[1])(24 × 0.15)                 | 3.60                | -                   |
|        | Suspended ceiling                                    | 0.18                | -                   |
|        | Imposed load(BS 6399[2]) Office loading              | -                   | 5.00                |
|        | (including partitions)                               | 3.78                | 5.00                |
| Walls  | Brick 105 mm thick (BS 648[1])                       | 2.16                | -                   |
|        | Blockwork 100 mm thick                               | 0.93                | -                   |
|        | Plaster 12 mm thick                                  | 0.22                | -                   |
|        |                                                      | 3.31                | -                   |

Beams   Concrete(24 × 0.15 × 0.3)          1.08 kN/m run
Columns Concrete (24 × 0.3 × 0.3)          2.16 kN/m run

*Load on Central Column*

Roof beams and column
    Beams (300 × 300)      2.16 × 6 × 2 =          25.92
    Column                 2.16 × 3.5 =            7.56
                                                   33.48

Second Floor
    Floor(due to area 2)   (3.78+5.00) × 6 × 6 = 316.08
    Beams per floor        1.08 × 6 × 2 =          12.96
                                                   329.04
First floor                as Second floor         329.04
ground floor (assuming suspended gnd floor as second floor)   329.04
Columns 1st-2nd and gnd-1st    2.16 × 7 =          15.12
                            Total Load    1035.72 kN

*Load on External Column*

Roof

Roof (spanning 12 m)  $(0.96 + 1.50) \times (12/2) \times 6 = 88.56$

Beams  $2.16 \times 6 \times 1.5 = \underline{19.44}$

108.00

Second Floor

Floor(due to area 1)  $(3.78+5.00) \times 6 \times 3.2 = 168.58$

Beams per floor  $1.08 \times 6 \times 1.5 = 9.72$

Column  $2.16 \times 3.5 = 7.56$

Wall  $3.31 \times 6 \times 3.5 \times 0.9^* = \underline{62.56}$

(* 0.9 takes account of window openings)  248.42

First floor  as Second floor  248.42

Ground floor  as Second floor  $\underline{248.42}$

Total Load  853.26 kN

## 6.5.2 Assessing the Ground

As we have seen in the previous section, there are different types of soil, all of which have different strengths and different allowable ground bearing pressures. When designing a foundation it is necessary to estimate an allowable ground bearing pressure for the soil on which the structure will sit and so it is important to identify the soil and test its strength. This is usually done by means of a 'soil investigation', consisting of a number of boreholes taken on site to a depth of 15 m or more during which in situ tests are carried out and samples taken for laboratory tests. The information found as a result of sinking a borehole is recorded in the form of a borehole log. A typical borehole log is shown in figure 6.33.

The log shows the layering (or stratification) of the soil, giving both depth and description of each layer. It shows that fill material exists to a depth of 1.7 m under which is a thin layer of sand and under this is a stiff natural clay. This information is correlated to the local Geological Maps showing drift deposits of Oxford Clay. Most importantly it shows the water table level and behaviour over time. This log shows that the water was found at a depth of two metres and the level rose by about 600 mm in 17 minutes which can indicate a low permeability of the surrounding strata. The log also shows that water was found at a depth of five metres and that the level remained constant for 20 minutes. The remarks section says that the water was sealed at three metres. Together this indicates that there is a perched water table in the fill above the natural clay and that the water table exists at five metres depth.

The log shows the results of a standard penetration test (S) at the depth 1.8-2.15 m in the form of an N number. This test is as described in the previous section and can be used to estimate allowable ground bearing pressure in gravels. Here the N value is 5 which, from table 6.2 with a width of footing of two metres, suggests an allowable ground bearing pressure of 40 kN/m² at a

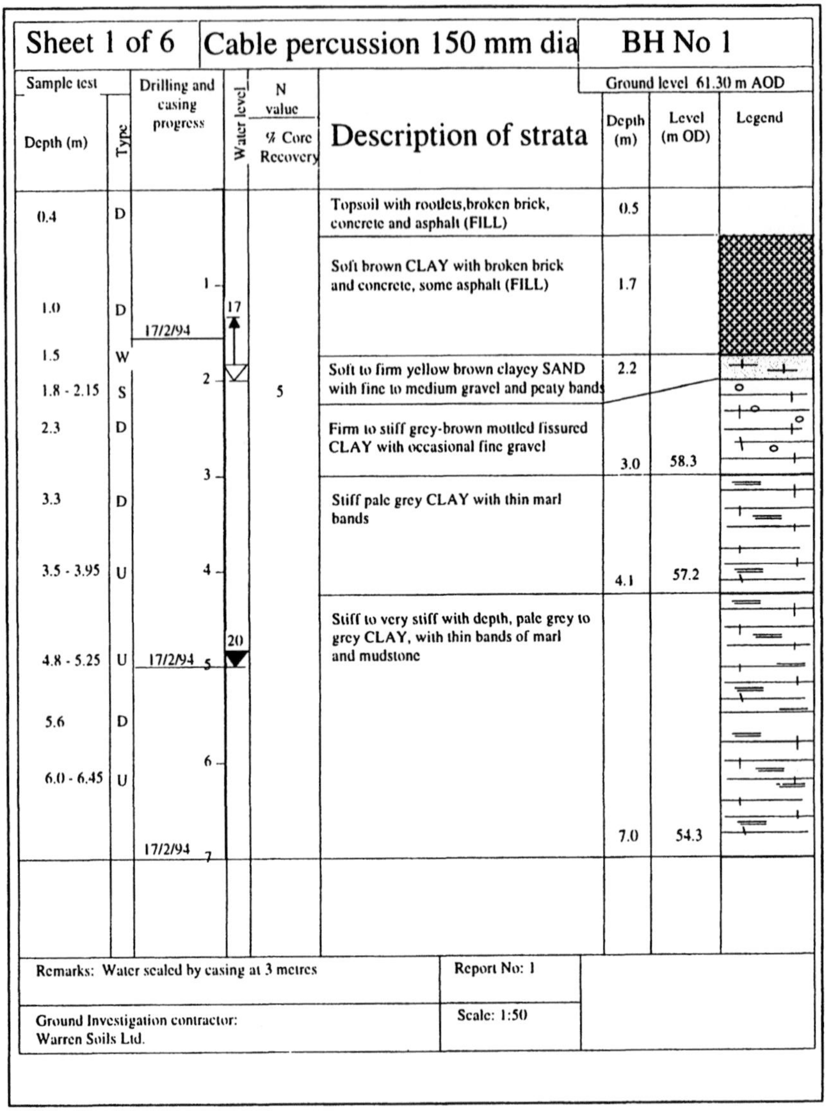

| Sheet 1 of 6 | | Cable percussion 150 mm dia | | | | BH No 1 | | |
|---|---|---|---|---|---|---|---|---|
| **Sample test** | | **Drilling and casing progress** | Water level | **N value** | **Description of strata** | Ground level 61.30 m AOD | | |
| Depth (m) | Type | | | % Core Recovery | | Depth (m) | Level (m OD) | Legend |
| 0.4 | D | | | | Topsoil with rootlets, broken brick, concrete and asphalt (FILL) | 0.5 | | |
| 1.0 | D | 1 — 17/2/94 | 17 | | Soft brown CLAY with broken brick and concrete, some asphalt (FILL) | 1.7 | | |
| 1.5 | W | | | | | | | |
| 1.8 - 2.15 | S | 2 — | | 5 | Soft to firm yellow brown clayey SAND with fine to medium gravel and peaty bands | 2.2 | | |
| 2.3 | D | | | | Firm to stiff grey-brown mottled fissured CLAY with occasional fine gravel | | | |
| | | 3 — | | | | 3.0 | 58.3 | |
| 3.3 | D | | | | Stiff pale grey CLAY with thin marl bands | | | |
| 3.5 - 3.95 | U | 4 — | | | | 4.1 | 57.2 | |
| 4.8 - 5.25 | U | 17/2/94  5 | 20 | | Stiff to very stiff with depth, pale grey to grey CLAY, with thin bands of marl and mudstone | | | |
| 5.6 | D | 6 — | | | | | | |
| 6.0 - 6.45 | U | | | | | | | |
| | | 17/2/94  7 | | | | 7.0 | 54.3 | |
| Remarks: Water sealed by casing at 3 metres | | | | | Report No: 1 | | | |
| Ground Investigation contractor: Warren Soils Ltd. | | | | | Scale: 1:50 | | | |

Figure 6.33

depth of two metres. Both disturbed (D) and undisturbed (U) samples have been taken and also a water sample (W). The disturbed samples are taken to aid description of the soil, to carry out moisture content tests and to determine the plasticity or grading. Such tests aid classification of the soil and are detailed in BS 1377[5]. Undisturbed samples are taken primarily to carry out tri-axial tests again described in BS 1377[5] to enable the estimation of cohesion and the value of internal angle of friction. As shown in the previous section it is the cohesion value which allows an estimate of ground bearing capacity in clays. Let us assume that the results of these tests indicate a 'C' value of 75 kN/m² at a depth of 3.5 and 4.8 m. The water sample is taken to test for water soluble sulphate content $SO_3$. If found to be 0.2 per cent or more, then sulphate resisting cement must be used for the concrete which will form the foundations. More details about sulphate exposure of concrete can be obtained from table 6.1 in BS 8110[6].

A number of such boreholes positioned across a site can give a three-dimensional underground picture of the soil profile enabling the Engineer to select the correct foundation type.

### 6.5.3 Designing the Foundation

Once the Engineer is satisfied that enough information is known about the ground on the site he can begin to consider the foundation options. It is considered easier to look at the following example to explain this process which, we recall, is shown in figure 6.31.

*Example*

For this simple example we shall consider the loads assessed in section 6.5.1 for the three-storey structure shown in figure 6.32 and the ground conditions shown in the borehole log figure 6.33. From the loads we can see that we must accommodate point loads of between 85.3 and 103.6 tonne. The soil profile indicates that an allowable bearing pressure of 40 kN/m² is possible on the sand at a depth of 1.8 m. For the purposes of this example we shall consider the design of a pad foundation.

$$\text{Area of pad required} = 1036/40 = 25.9 \text{ m}^2$$

$$\text{Size of pad required} = \sqrt{25.9} = 5.09 \text{ m square}$$

There are a number of problems associated with this solution. The first is that the pad size is greater than the two metres assumed in table 6.2. For a four metre wide footing, table 6.2 suggest an ABP of about 30 kN/m². This ABP gives rise to an even wider footing which will be uneconomic, clash with neighbouring footings, and may not even then achieve the required load capacity. Secondly, the presence of the perched water table would mean that the construction of a pad footing would be hampered by the ingress of water which may soften the formation. Thirdly, the settlement characteristics would be determined by the underlying clay due to the thin layer of sand. In this case the best solution is a piled foundation using the stiff clay layers as the main support.

The problem with this solution is that the piles will almost certainly go deeper than seven metres, the current depth of investigation. If piles are to be used then the Engineer would have to go back to the Client and ask for further, deeper soil investigations. For the purpose of this example, however, we shall assume that

the stiff clay extends down below seven metres for a sufficient depth to be able to use it for our foundations. The C values have been assessed by tri-axial tests at 75 kN/m² and we shall assume this value for the full length of the designed pile. The pile is predominantly a friction bearing pile and we shall choose a diameter of 350 mm. Clay begins 2.2 m from the surface and so for a first try consider a pile length of 12.2 m, i.e. an effective length is 10 m.
Using Skempton's formula:

$$Q_a = \frac{\alpha P l c + 9 A_b c}{\text{Factor of safety}}$$

$$Q_a = \frac{0.45 \times (2 \times \pi \times 0.175) \times 10 \times 75 + 9 \times (\pi \times 0.175^2) \times 75}{2.5}$$

$$Q_a = \frac{371 + 65.0}{2.5} = 174.4 \text{ kN}$$

Thus we would need 5 piles under the outside column and 7 under the central column. This is not satisfactory since a 5 pile arrangement is the usual maximum economic arrangement; this must be checked with the specialist sub-contractor. Try 17.2 m long piles:

$$Q_a = \frac{0.45 \times (2 \times \pi \times 0.175) \times 15 \times 75 + 9 \times (\pi \times 0.175^2) \times 75}{2.5}$$

$$Q_a = \frac{556.7 + 65.0}{2.5} = 248.6 \text{ kN say } 249 \text{ kN}$$

Now we find that we need 4 piles under the edge columns and 5 piles under the centre column. This is enough to allow for some 10 to 12 per cent reduction in the working load of the pile due to the bulb effect. Whether the bulb effect reduction factor is taken into account or not depends on the judgement of the Engineer and specialist piling sub-contractor. If the full 30 per cent is taken the piles will need to be 19.2 m long.
    The construction of the piles will probably be continuous flight auger type in situ concrete piles, using sulphate resisting concrete. The piles will probably need a shell for the first two metres to protect against ingress of the perched water table and to stabilise the made ground. A permanent plastic sleeve is also recommended for the first two metres to protect against negative skin friction.

## 6.6 GROUND IMPROVEMENT

Due to a shortage of prime development land it is becoming more common for Engineers and developers to attempt to improve the load bearing capacity of existing poor ground conditions and develop land which would have previously not been considered usable. The process of ground improvement relies on compaction and consolidation to increase density, shear strength and load bearing performance. We have already made reference to some ground water control techniques which in themselves would improve allowable ground bearing pressure, i.e. vertical drains and electro-osmosis (see chapter 4). The following ground improvement techniques have been developed over the past 25 years and principally rely on compaction by some mechanical means. These include:

- Dynamic compaction
- Vibro compaction
- Vibro replacement
- Vibro flotation
- Jet grouting
- Pressure grouting

It may be that a combination of two or more of the above techniques can be used to achieve the desired effect of increasing the allowable ground bearing pressure and reduce settlement.

### 6.6.1 Dynamic Compaction

This process simply compacts the ground by dropping a heavy weight of between 10 and 20 tonne a height of between 15 to 25 m on to the soil. The weight is usually two to three metres in diameter and so leaves a crater up to five metres diameter and 1.5 m deep, depending on the stiffness of the ground. Vertical drains (see chapter 4) are sometimes installed before compaction begins to allow rapid draining of excess ground water driven out of the soil by the compaction process. The centres of impact are again a function of the stiffness of the ground and the degree of compaction required. Figure 6.34 shows a typical arrangement of the first and second drop pattern on plan. Compaction is usually accompanied by a general reduction in ground level.

### 6.6.2 Vibro Compaction

Vibro compaction uses a large vibrating poker to agitate loose soil into a more compact and dense arrangement, see figure 6.35. The poker is 300 to 450 mm in diameter and weighs some 2 to 4 tonne; it is vibrated by compressed air which rotates an eccentric weight inside the tip of the poker. The frequency and power of the vibrations are such that it causes the soil to 'liquefy' on contact; the poker can thus be lowered into the ground and be allowed to penetrate under its own weight. The soil will collapse, filling in the hole left by the poker as it is withdrawn. The poker is repeatedly inserted and withdrawn in decreasing depths to allow full compaction of the soil which fills the hole. By adding sections to the poker, it can be inserted to a depth of between 5 and 15 m. The type of soil best suited to this type of compaction is loose sand and gravel. A general reduction in ground level is expected with this compaction.

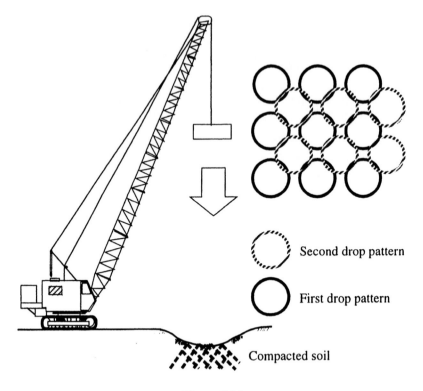

Second drop pattern

First drop pattern

Compacted soil

Figure 6.34

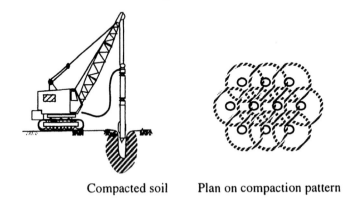

Compacted soil        Plan on compaction pattern

Figure 6.35

### 6.6.3  Vibro Replacement

This process is very similar to vibro compaction except that here an additional
stone fill is poured into the hole during compaction. This results in a compacted

stone column formed in the ground where the poker penetrated. The diameter of the column depends upon the stiffness of the soil and will be at least the diameter of the poker and can be up to one metre in diameter. The repeated insertion and withdrawal of the poker allows more stone to be introduced into the hole and compacts it radially outwards into the soil giving rise to a compacted zone, see figure 6.36.

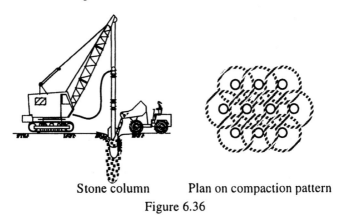

Stone column         Plan on compaction pattern

Figure 6.36

This is the most popular form of ground improvement in this country today because it is cheap, effective, tried and tested. The area for compaction need only be the area directly under the foundations to be supported and the spacing of vibro points is anything between 0.6 to 3 m. For example a weak soil of ABP of 50 kN/m² can be improved to 100 kN/m² by using vibro replacement at 1.2 m centres. If a ground bearing floor slab is to be used then a general area can be treated at wider centres. In terms of quality control each vibro point can be located before the construction of the foundation, thus ensuring that the area beneath an intended foundation has indeed been treated. Also a California Bearing Ratio (CBR) test can be carried out. The CBR test gives a measure of the stiffness and thus load bearing qualities of the soil. See chapter 10.

### 6.6.4  Vibro Flotation

This is the same process as used in vibro replacement, except that high pressure water jets at the tip of the poker achieve initial penetration of the soil. The vibrating action is later used to compact the stone column. This process has the advantage of penetrating hard layers of ground in combination with a high water table, but it is dirty work and imprecise and is not often used in the UK.

### 6.6.5  Jet Grouting

This is a cost-effective method of improving almost any soil from gravel to clay irrespective of ground water conditions. The process is shown in figure 6.37. First a steel pressure tube is drilled into the ground to the required depth. As it is withdrawn the tube is slowly rotated while a cement grout is jetted out sideways at great pressure, up to 70,000 N/mm². The jetting action either injects the grout into the fabric of the soil, or it displaces the soil sideways, replacing it with the grout depending upon the soil type and grain size. The result is a cemented column of soil which if installed at regular spacing has the effect of improving the stiffness of the soil. Such a system also has the added advántage of producing an impermeable barrier against ground water and so at sufficiently close centres it can be used for ground water control. Similar spacings as for vibro replacement can have similar beneficial effects on the soil.

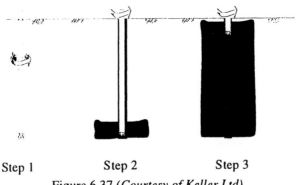

Step 1            Step 2            Step 3

Figure 6.37 (*Courtesy of Keller Ltd*)

### 6.6.6 Pressure Grouting

This also requires a steel pressure tube drilled into the ground to the required depth. Grout is forced down the tube and out at the end to fill any voids or cracks in the soil or rock with grout.

### 6.7 UNDERPINNING

The foundation arrangements considered so far have been constructed on the same level, i.e. the contact pressure is applied at the same level. This is not always the case and there are many occasions when the Engineer will be faced with the problem of building a new foundation adjacent to and lower than existing foundations, see figure 6.38. It is also necessary on occasion to underpin existing foundations where for some reason the ground has become unstable. This section looks at the various options to cope with the above situations.

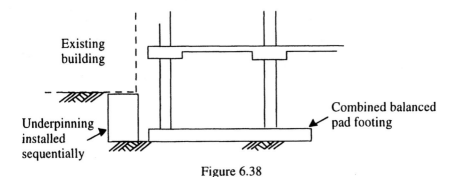

Existing
building

Underpinning
installed
sequentially

Combined balanced
pad footing

Figure 6.38

The most common form of underpinning is that used for existing brick
bearing structures where the foundations have suffered movement due to clay
shrinkage, notable in the South East of England. The idea is that concrete is
placed below the wall down to a level where the ground is more reliable. For a
cellular structure, i.e. continuous load bearing brick or block walls, the
underpinning is achieved by sequentially digging out by hand small lengths of
wall and concreting them before digging the next section. This method is shown
in figure 6.39. Sections numbered 1 are dug out and concreted, then sections
numbered 2 and then sections numbered 3. The result is a fully underpinned
wall with concrete extending down to firm ground. Each section is packed with
a 'dry' mortar mix between the new concrete and the underside of the original
footing to achieve a good structural support.

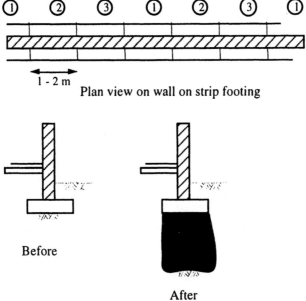

① ② ③ ① ② ③ ①

1 - 2 m

Plan view on wall on strip footing

Before

After

Figure 6.39

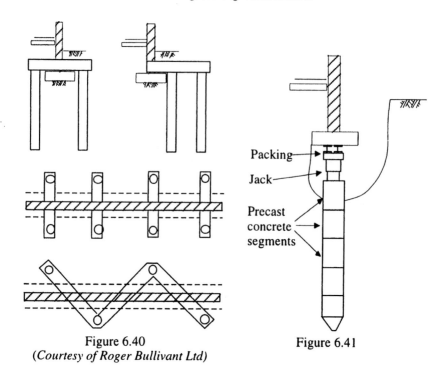

Figure 6.40
(*Courtesy of Roger Bullivant Ltd*)

Figure 6.41

### 6.7.1 Needle and Piles

Of course, it may not be possible to reach reliable ground by a hand-dug method, in which case piling can be used in underpinning. There are three methods of underpinning commonly used, needle and pile, jack piles and drilled piles.

The needle and pile system uses conventional piles drilled either side of the existing wall, between which spans a ground beam called a 'needle' which supports the wall. The arrangement is shown in figure 6.40. If access cannot be gained to one side of the wall, then a cantilever needle system can be used, also shown in figure 6.40.

### 6.7.2 Jack Piles

Jack piles are piles which are installed directly under the wall. They are jacked into position against the self-weight of the wall by using hydraulic jacks. The piles used are precast in small sections so that they can be installed in a confined space. See figure 6.41.

### 6.7.3 Drilled Piles

Called Palo Radice, the system was first developed in Italy about 25 years ago to underpin the subsiding structures of Venice. The method, shown in figure 6.42, uses a tungsten tipped drill tube to drill straight through the existing foundations and on into the supporting ground. Water or bentonite is used to flush out the tube and a high strength grout is pumped into the tube to form the pile. Reinforcement is used full length and can be a single bar for small diameters or a full reinforcement cage for larger diameters. (Indeed the tube can be used as reinforcement to provide moment resistance if required.) Whilst the concrete is in liquid form the concrete is continually topped up as the tube is withdrawn. For a bridge this system has the advantage of being able to install piles from the bridge deck, drilling straight through the abutments and piers into the soil.

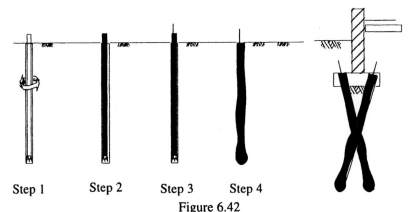

Step 1        Step 2        Step 3        Step 4

Figure 6.42

*(Courtesy of Fondedile Foundations Ltd and Roger Bullivant Ltd)*

### 6.7.4 Framed Structures

For framed structures the structure must be temporarily supported with jacks and/or props whilst the new base is constructed.

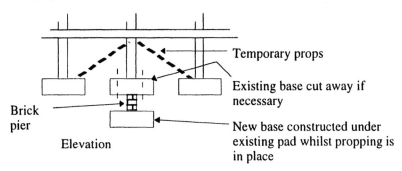

Temporary props

Existing base cut away if necessary

Brick pier

New base constructed under existing pad whilst propping is in place

Elevation

Figure 6.43

Of course, all these schemes must be correctly designed. The existing bases must be checked to see if they can carry the increase in vertical load and some form of horizontal restraint is required between the two bases. The bracing props must also take a considerable compressive load.

## 6.8 REFERENCES

1.  British Standards Institute. *BS 648: Schedule of Weights of Building Materials*, London 1964.
2.  British Standards Institute. *BS 6399: Design Loading for Buildings, part 1: Code of Practice for Dead and Imposed loads,* London 1984.
3.  British Standards Institute. *CP 3: Chapter V: Loading, Part 2: Wind Loads,* London 1972.
4.  British Standards Institute. *BS 8004: British Standard Code of Practice for Foundations.* London 1986
5.  British Standards Institute. *BS 1377: Part 1 to 9: 1990: British Standard Method of Test for Soils for Civil Engineering Purposes.*
6.  British Standards Institute. *BS 8110: Part 1 to 3: 1985: Structural Use of Concrete.*
7.  G. Barnes, *Soil Mechanics,* 1995 Macmillan Press ISBN 0-333-59654-4
8.  Jürgenson, L. *The Application of Theories of Elasticity and Plasticity to Foundation Problems,* Proc. American Society of Civil Engineers: 1934

# 7 Retaining Walls and Deep Basements

Much of the work of a Civil Engineer involves opening the ground in a safe way to construct underground structures. Retaining walls can be used to provide support to the ground in both the permanent and temporary condition, but they are expensive to construct, so the Engineer will try to use an embankment or self-supporting slope wherever possible. Only after eliminating this option will a retaining structure of some form be considered. Retaining walls are designed to provide support to otherwise unstable soil surfaces and are often an integral part of basement design. In particular, deep basements may incorporate several storeys of a structure below ground and must be designed with care to resist very large forces from the ground. In this chapter we shall consider the construction and design of retaining walls and deep basements. Looking initially at conventional retaining walls, we shall move on to discuss new developments in this field of construction linking them in to basement and deep basement construction.

## 7.1 CONSTRUCTION OF RETAINING WALLS

Retaining walls are required where the design dictates that a rapid change in ground level is called for in a small area and the ground is not strong enough to stand on its own. There are two main categories of retaining wall, gravity and cantilever. Within these categories are different types and these are summarised in table 7.1.

Table 7.1

| Category | Type |
|----------|------|
| Gravity | Mass Concrete |
| | Crib walls or gabions |
| | Reinforced earth |
| Cantilever | Concrete (counterfort and buttressed) |
| | Steel (see section 4.2.2) |
| | Bored pile (see section 4.3.5) |
| | Diaphragm (see section 4.3.2) |
| | Anchored |

### 7.1.1  Gravity Retaining Walls

Gravity retaining walls are used where a small slope is allowed to the retained face and there is enough land available to accommodate the thickness of the construction, which can be equal to the retained height, see figure 7.1. They are used to retain soil, gravel or even waste material, but are not considered to be water retaining unless special design features are incorporated. They are often used for retaining embankments in road construction.

Gravity walls depend on their weight to retain soil. They are designed to distribute their self-weight, sometimes called 'dead weight' and the forces from the retained soil to the ground on which it sits without undue settlement occurring. Mass concrete retaining walls are limited in height to about three metres as they become uneconomic at greater heights. Generally, the width of the base is about one third of the height and at the top one sixth, as shown in figure 7.1(a).

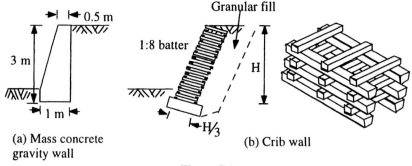

(a) Mass concrete gravity wall

(b) Crib wall

Figure 7.1

*Crib Walls*

This wall is shown in figure 7.1(b) and is built up of small concrete or hardwood beams stacked in a grillage pattern. They are designed to interlock and are filled with a granular material. The granular material usually extends up to one metre behind the wall. The principle of operation is that the beams act as an interlocking agent with each beam supporting a small amount of fill. The combined action is for the fill to act as one stable solid mass and act as a gravity retaining wall; topsoil can be placed within the grillage on the outer face allowing vegetation to grow and enhance the appearance.

*Gabions*

Gabions are wire mesh boxes filled with well-packed stones 100 mm to 225 mm in size. The boxes are generally one metre wide by one metre deep and two metres long, and are placed on top of each other in a brickwork pattern (English bond) to form a wall. The result is a heavy construction acting as a gravity retaining wall.

*Reinforced Earth*

This method combines the strength of the reinforcing material with that of the soil to produce a stable bulk capable of retaining soil. The construction consists of about 250 mm thick, well compacted layers of soil between reinforcing mesh or strips of reinforcement. The mesh may be of steel or a high tensile plastic, usually polypropylene, and is attached to precast concrete facing panels to give a neat finish. This type of construction is used widely because of its speed and low cost in comparison to other forms of construction. A reinforced earth wall is only about half the weight of a conventional concrete retaining wall and so can reduce the contact pressure which can be an advantage in poor ground. This type of wall cannot be expected to retain liquids without additional design measures, but it is commonly used to retain earth embankments and bridge abutments in road construction.

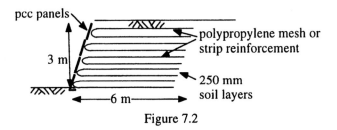

Figure 7.2

The height is limited only by the amount of room available for reinforcement which must extend back at least a distance equal to twice the height of the retained soil and in some cases further.

## 7.1.2 Cantilever Retaining Walls

These are so described because the main structural member, the stem, which retains the soil, cantilevers as a beam from the base or foundation. The concrete cantilever wall uses the weight of the soil retained to contribute to stability against overturning and sliding, see figure 7.3.

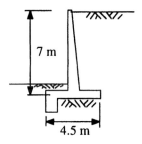

Figure 7.3 Concrete cantilever retaining wall

Concrete cantilever walls are often used to retain liquid, although they are also used to retain soil. These walls can also resist heavy vertical loads and are often incorporated into bridge abutments where they can be used to support the bridge deck. Concrete retaining walls can retain a height of seven metres without any other means of support. At this height the general proportions are as shown in figure 7.3, assuming a granular backfill and an allowable ground bearing pressure of 150 kN/m². Wall heights in excess of seven metres can be achieved by using counterforts or buttressing, see figure 7.4.

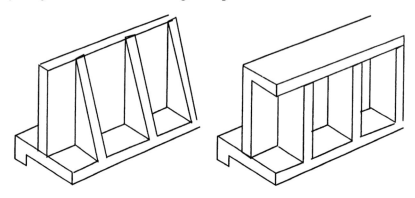

Figure 7.4 (a) Counterfort wall    Figure 7.4 (b) Buttressed wall

*Steel Cantilever Retaining Walls*

We have previously discussed steel sheet piling as a form of temporary ground water control; see chapter 4, section 4.2.2, where the installation and design of steel piles are looked at in detail. However, steel sheet piling is also used to retain soil in the permanent and temporary condition. Steel sheet piling is very useful in situations where the retained soil cannot be disturbed or where space is limited.

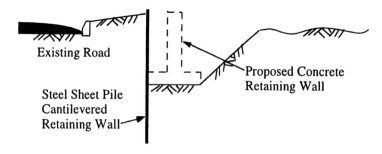

Existing Road

Steel Sheet Pile
Cantilevered
Retaining Wall

Proposed Concrete
Retaining Wall

Figure 7.5

For example, where an excavation is to be carried out immediately adjacent to a road, steel sheet piles can be used in a temporary condition to allow the construction of a concrete retaining wall as shown in figure 7.5. Steel is not recommended in the permanent situation because it is too flexible and steel would be vulnerable to rusting from ground water and de-icing salts.

In the above example the piles are installed in a cantilever mode, but to improve performance a propping or ground anchor system can be used. A propping arrangement is shown in figure 7.6.

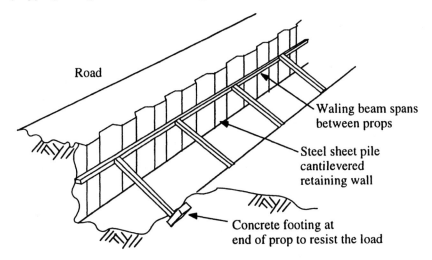

Road

Waling beam spans between props

Steel sheet pile cantilevered retaining wall

Concrete footing at end of prop to resist the load

Figure 7.6

Props are spaced at between two and six metre centres depending upon ground conditions and the size of waling beam. Propping not only reduces deflection under load at the top of the piles but can also reduce the length of the pile which extends below the excavation, thus saving money. The disadvantages of this system are that the props restrict access to the excavation, can slow construction and are vulnerable to accidental damage. Ground anchors overcome these problems. They consist of steel bars drilled into the ground at centres as shown in figure 7.7. The anchors are usually spaced at two to three metre centres but can vary, depending upon ground conditions, the size of pile and the size of the waling beam. The advantages offered by ground anchors are the same as for propping but they do not restrict access and if extended far enough they can protect against slip circle failure, (see later). The disadvantages of this system are that the anchor may extend under adjoining properties and could potentially produce ground movement. This of course can be designed against, but the legal complications of getting permission to extend work beyond the site boundary can turn the Engineer's mind to a cantilever or a propping system. The fixing provided by ground anchors is a mechanical fixing into the soil provided by either grouting, screwed or auger ends, the purpose being to mobilise passive resistance of the soil which is used to stabilise the piles.

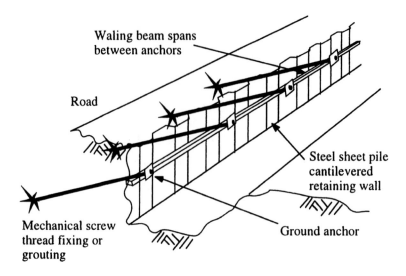

Waling beam spans
between anchors

Road

Steel sheet pile
cantilevered
retaining wall

Mechanical screw
thread fixing or
grouting

Ground anchor

Figure 7.7

*Contiguous, Secant and Diaphragm Walls*

Once again, the construction of these types of wall is discussed in detail in
chapter 4 in the context of ground water control; but these walls are also used
extensively in earth retaining situations. The system is similar to that considered
above except that the steel sheet piles are replaced with a concrete wall of some
form. Figure 7.8 shows a contiguous wall using a ground anchor system for
support.

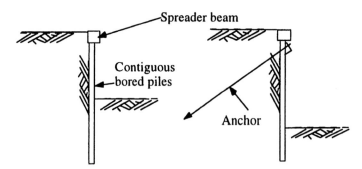

Spreader beam

Contiguous
bored piles

Anchor

Figure 7.8

Note that a spreader beam is used in this case. This is because there is no
waling to span between the anchors and there must be some structural member
supporting piles in between the ground anchors spreading possible load
concentrations.

## 7.2 DESIGN OF A RETAINING WALL

A retaining wall is designed to resist forces from ground and water pressure, but in this case we shall consider ground pressures alone, assuming that any ground water is removed by drainage provided behind the wall. As we have seen in section 4.2.3, the sideways pressure from the ground is called the 'active pressure' given by Rankine's formula and we must consider this as our loading on to the structure.

$$\text{Pressure} = K \times \gamma \times H \quad (kN/m^2)$$

Where  $K$ = Rankine's constant
$\gamma$ = Density of the soil
$H$ = Depth below the surface.

The active pressure distributions are shown in figure 7.9 and the resulting forces applied by the ground are:

$$P_a = K_a\, \gamma \frac{H^2}{2} \quad kN/m \text{ run of wall}$$

$$P_p = K_p\, \gamma \frac{D^2}{2} \quad kN/m \text{ run of wall}$$

Where $K_a$ is Rankine's constant for active pressure, usually 0.33 for granular soil $K_p$ is Rankine's constant for passive pressure approximately 3.0 for granular soil.

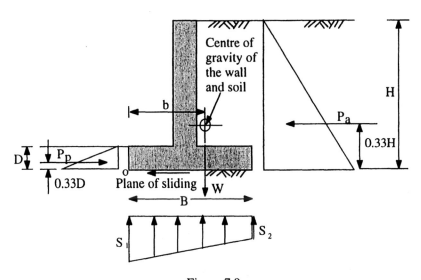

Figure 7.9

With these forces in mind we can now look at the design method. All retaining structures are designed to resist active pressures and we must consider the following factors:

- Stability against overturning
- Allowable ground bearing pressures
- Stability against sliding
- Circular slip
- Material strength of the wall

We shall consider each in turn in some detail, with examples.

### 7.2.1 Stability Against Overturning

Here we must consider all the active forces that are trying to push the wall over and make sure that they are outweighed by the passive forces holding it up. We do this by taking moments about 'o' (as shown in figure 7.9) of all the forces acting on the wall. The active moments must be outweighed by the passive moments by at least a factor of 2, which will mean that the wall has a factor of safety against overturning of two.

$$\text{Factor of safety} = \frac{Wb + P_p 0.33D}{P_a 0.33H} = 2.0 \text{ minimum.}$$

Where W is the weight of the retaining wall and the retained soil.

### 7.2.2 Allowable Ground Bearing Pressures

This calculation is carried out in a similar manner to that considered in foundation design (section 6.3.1), checking that e is less than a third of B so that there is no uplift.

First we need to calculate the overturning moment $= P_a 0.33H$

Then find the eccentricity e from the equation $P_a 0.33H = We$

$$\text{when} \quad \frac{We}{W} < \frac{B}{6} \quad S_1 = \frac{W}{BL} + \frac{We}{Z} \quad S_2 = \frac{W}{BL} - \frac{We}{Z}$$

Where  We  =  Overturning moment
       W   =  Total weight of wall and retained soil
       B   =  width of base
       L   =  length of base (This will be one metre if we consider the wall per metre run)
       Z   =  section modulus $= \dfrac{LB^2}{6}$

### 7.2.3 Stability Against Sliding

To design against sliding instability we must ensure that sufficient friction is mobilised on the underside of the base to be at least twice the value of the active force $P_a$.

$$\text{Thus Factor of safety} = \frac{P_s + P_p}{P_a} = 2.0 \text{ minimum.}$$

Where $P_s$ = force mobilised by friction = $W \tan \phi$
(Where $\phi$ is the internal angle of friction.)

### 7.2.4 Circular Slip

This type of failure is common in weak cohesive soils and is shown in figure 7.10. The analysis is similar to that carried out for slope stability in section 5.4.2.

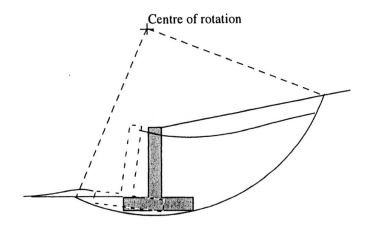

Figure 7.10

Special consideration must be given to weathering, ground water movement and the removal of soil weight at the toe of the retaining wall. Both weathering and ground water movement can reduce the frictional strength of the soil. In addition an initially high ground water table may increase the weight of the retained soil, irrespective of whether the water provides hydrostatic pressure on the back of the wall. Removal of soil at the toe of the wall will reduce the stability of the slip circle and this must be compensated for by either frictional resistance of the remaining slip circle, with a factor of safety of 2, or increased weight of the wall. If extra weight can be applied to the left hand side of the centre of rotation, then the effect will be to increase the stabilising moments.

### 7.2.5  Material Strength of the Wall

Both the stem and base of the retaining wall must be designed to resist the forces
applied to them from the ground. The concrete will need reinforcement designed
to either BS 8110[2] in the general case, or BS 8102[1] if it is to retain liquid. The
base can be designed as a cantilever spanning from the wall stem in both
directions and will probably require the main steel to be placed at the bottom, for
the toe, and the top for the heel. The stem is designed as a cantilever spanning
from the base.

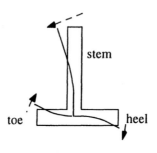

Figure 7.11

### 7.2.6  Example

Consider the wall shown in figure 7.12 to retain granular soil of density 18
kN/m³, internal angle of friction $\phi$ of 30° which gives $K_a = 0.3$ and $K_p = 3.0$ and
formation of the same soil with an allowable bearing pressure of 200 kN/m².

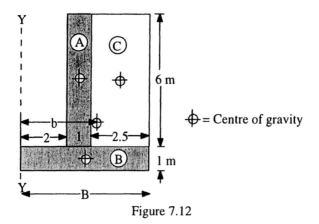

Figure 7.12

*Loading*

$$P_a = K_a \, \gamma \frac{H^2}{2} = 0.33 \times 18 \times \frac{7^2}{2} = 145.5 \text{ kN/m run of wall}$$

$$P_p = K_p \, \gamma \frac{D^2}{2} = 3.0 \times 18 \times \frac{1^2}{2} = 27 \text{ kN/m run of wall}$$

*Stability Against Overturning*

The position of the centre of gravity of the wall and the retained soil 'b' can be found by taking first moment of area about the axis Y - Y as shown in figure 7.12.

$$b = \frac{2.5 \times 6 \times 1 \times 24 + 4.25 \times 6 \times 2.5 \times 18 + 2.75 \times 1 \times 5.5 \times 24}{6 \times 1 \times 24 + 6 \times 2.5 \times 18 + 1 \times 5.5 \times 24}$$

$$b = \frac{360 + 1147.5 + 363}{546} = 3.426 \text{ m} \qquad (W = 546 \text{ kN})$$

$$\text{Factor of safety} = \frac{Wb + P_p 0.33D}{P_a 0.33H} = \frac{546 \times 3.426 + 27 \times 0.33 \times 1}{145.5 \times 0.33 \times 7} = 5.6$$

Factor of safety = 5.6 > 2, therefore stability against overturning is sufficient.

*Allowable Ground Bearing Pressures*

Rearranging the equation from section 7.2.2   $P_a 0.33H = We$

$$e = \frac{P_a 0.33H}{W} = \frac{336.1}{546} = 0.616$$

$$\frac{B}{6} = \frac{5.5}{6} = 0.917 > 0.616 \text{ therefore uplift will not occur.}$$

$$Z = \text{section modulus} = \frac{LB^2}{6} = \frac{1 \times 5.5^2}{6} = 5.042 \text{ m}^3/\text{m run}$$

$$S_1 = \frac{W}{BL} + \frac{We}{Z} = \frac{546}{1 \times 5.5} + \frac{336.1}{5.042} = 166 \text{ kN/m}^2$$

$$S_2 = 32.6 \text{ kN/m}^2$$

These ground pressures are within the allowable, but the variation is large. This is acceptable in a gravel formation, but may not be in clay, due to long term differential settlement.

*Stability Against Sliding*

$P_s$ = Force mobilised by friction = W tan $\phi$ = 546 × tan 30° = 315.3 kN

Factor of safety = $\dfrac{P_s + P_p}{P_a} = \dfrac{315.3 + 27}{145.5} = 2.4 > 2$       therefore acceptable

## 7.3 DEEP BASEMENTS

This type of construction is used on confined sites where land prices are high such as in London. Deep basement design can be complex and is too specialised to consider here in any depth, but a general background knowledge of the problems involved is appropriate at this stage. In the design we must consider the following loading conditions:

- maximum load from building above the basement
- minimum load from building above with upward water pressure and heave
- active forces from the surrounding soil onto the perimeter walls.

### 7.3.1 Maximum Load from Building Above

Deep basements are usually under high structures of 20 to 30 floors. Such tall structures give rise to high loads typically about 3000 kN per column. Support of such loads in a confined space can be accommodated by the use of large diameter piles. Piles of 1200-2000 mm diameter are bored into the ground to depths of 50-60 m with the ends of the piles spread out in a bell shape, called under-reamed, to spread the load, as shown in figure 7.13. These piles are designed in a similar way to the method shown in section 6.4.1. This type of pile is utilising a large proportion of end bearing capacity to achieve support of the high loads. Alternatively, concrete pads can be cast monolithically with the ground floor slab to produce a raft construction. The thickness of the pads and the floor are approximately 1.5 m and 0.5 m respectively. The majority of the load is resisted by the pad footing but as this settles the floor also begins to resist load. Such loads are called 'shade off pressures' and must be designed for. The effect of this is to cause the floor to span between columns for which steel reinforcement must be provided.

### 7.3.2 Minimum Load from the Building

This condition must be considered because the uplift due to hydrostatic pressure and heave can be quite considerable. In some case anchor piles may be required, as shown in figure 7.14. In an effort to avoid the forces due to heave or ground water the ground slab may be cast separately from the main load bearing structure and allowed to 'float' as shown in figure 7.15, although this cannot deal with ground water seepage. Alternatively, a suspended ground slab may be considered as shown in figure 7.16.

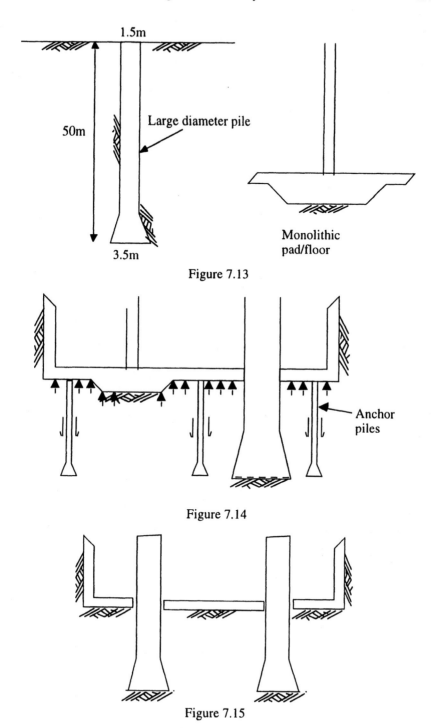

1.5m

50m

Large diameter pile

3.5m

Monolithic
pad/floor

Figure 7.13

Anchor
piles

Figure 7.14

Figure 7.15

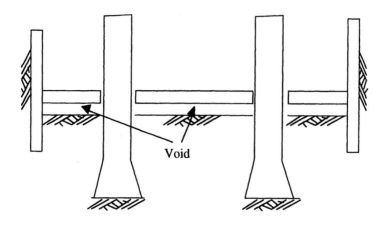

Figure 7.16

Here the void must be vented and drained. If there is a problem of variable water levels it may be difficult to waterproof this construction.

### 7.3.3 Active Forces from the Surrounding Soil

These forces come from ground pressure and water pressure. They may be resisted by conventional concrete retaining walls constructed in an open cast situation, or with temporary sheet piling, as shown in fig 7.17.

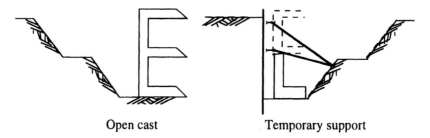

Open cast          Temporary support

Figure 7.17

These forces may also be resisted by bored piling, steel sheet piling or a diaphragm wall in either cantilever mode or with ground anchors. All these methods are limited in depth; top-down construction (see section 7.3.4) overcomes this. Advantages and disadvantages of a cantilever structure as compared to a ground anchor structure are shown in figure 7.18.

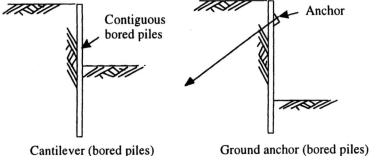

Cantilever (bored piles)                    Ground anchor (bored piles)

*Disadvantage:*                              *Disadvantage:*
• Care needed against movement              • Movement of ground around
                                              anchors.
• Long piles needed

*Advantage:*                                 *Advantage:*
• No Propping                                • No Propping

Figure 7.18

## 7.3.4 Top-Down Construction

This technique is the most widely used form of construction for deep basements.
It solves most of the problems associated with conventional retaining wall
construction. Here work may continue above and below ground at the same
time. The process is shown in figures 7.19, 7.20 and 7.21.

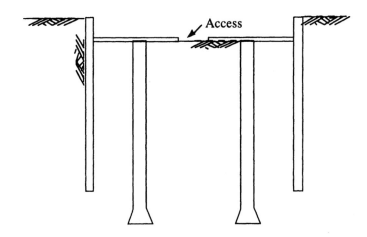

Figure 7.19

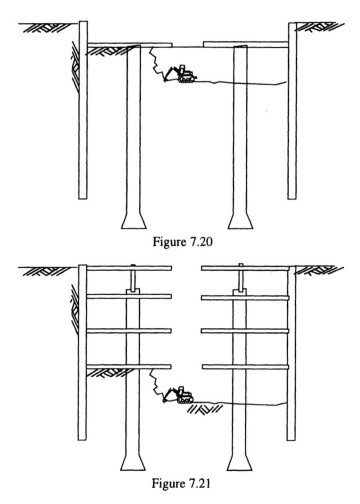

Figure 7.20

Figure 7.21

First, the large-diameter piles are drilled and concreted in the positions required to support the main structural columns of the upper structure. A contiguous, secant piled or diaphragm wall is used around the perimeter of the basement. Such piling can be installed in depth of up to 50 m. Such construction causes little disturbance to the ground and is easy to install and so can be used close to existing buildings. The ground is levelled and prepared to receive the concrete that will form the ground level floor. A hole approximately ten metres square is left unconcreted in the centre of the slab to provide access for excavation plant. When the slab is complete soil is excavated from underneath to form the first floor basement. Again the ground is prepared to receive the concrete which will form the first basement floor and an access is left in the middle for excavation plant. Each concrete slab is keyed into the perimeter wall and the large diameter piles to gain support. Naturally, drop beams must be cast within the floor to support it and large diameter piles can be replaced by steel columns if extra floor space is needed. The exposed perimeter wall is treated with water proofing and clad. The main advantage of this system is that work can commence on the superstructure before the basement is complete, cutting overall construction time and considerably reducing costs. Other advantages are that temporary internal strutting is eliminated and all work carried out is permanent work so the Contractor need not spend time and money on temporary works. Ground disturbance outside the basement is minimal, thus reducing the risk of settlement damage to adjacent buildings. One disadvantage is that excavation under slab can be difficult and costly. Another disadvantage is that ground water control measures can be difficult to install and the watertight integrity of the external wall cannot be guaranteed. For further information and case studies of deep basement construction, see Design and Construction of Deep Basements[4]

## 7.4 WATERTIGHT BASEMENTS

Watertight sub-structures are constructed in accordance with BS 8102[1]. This standard defines levels of water protection, from Grade 1 which allows some water penetration to Grade 4, which does not allow water penetration and demands a totally controlled environment. It is of absolute importance to the success of the design of a basement that the Engineer has a clear understanding of the Client's requirements and that the Client has a clear understanding of the limitations of the construction. It is impractical for any sub-structure to be 100 per cent waterproof, firstly due to the high standards of workmanship required in a difficult construction environment and secondly due to the costs which would be very high. The answer is to provide the Client with an underground environment that matches his needs. To provide water protection the Engineer has three basic types of construction at his disposal: tanked, monolithic and drained cavity construction. These are shown in table 7.2 together with the classification types used by BS 8102[1]. We shall look at each of these types of construction in more detail before we discuss how they can be used to attain the class of protection required by the Client.

### 7.4.1 Tanked Protection (Membranes)

Tanking is an impermeable barrier included somewhere within the construction of the basement walls and floor, as shown in figure 7.22. Placed on either internal or external faces of the structure, water protection relies entirely on the integrity of the membrane. Traditional membrane construction was of asphalt 20 mm thick and applied in a similar way to that of plaster on a brick wall. The problem with this system is that the asphalt has to be applied hot and as such it is a slow and expensive process. More recently proprietary systems have become commercially available which are a combination of bitumen-coated polythene sheets which can literally be stuck onto the surface that needs protection and which may be used in lieu of asphalt. Applied externally, adhesion is assisted by the active pressure of the soil retained, but it the membrane will require some sort of protection against abrasion from the backfilling. Applied internally, the membrane may be forced away from the wall by water pressure and in this situation an internal wall may be used to keep it in place. In the internal position the membrane provides no protection for the external structural wall against aggressive ground conditions such as sulphates.

Table 7.2

| Construction | Structure Classification in accordance with BS 8102 |
|---|---|
| Tanked protection (membranes) | Type A |
| Monolithic structures | Type B |
| Drained cavities | Type C |

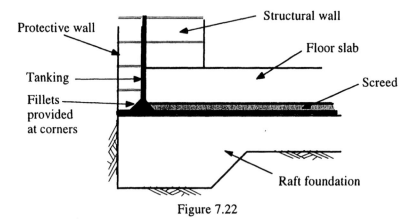

Figure 7.22

### 7.4.2 Monolithic Structures

Water protection can be provided to some extent by the use of a solid concrete wall. These structures rely on the quality and density of the concrete to act as a barrier to moisture penetration. Such walls can be designed to BS 8110[2] to minimise water penetration or BS 8007[3] to prevent water penetration. All concrete when set and cured will have some degree of cracking present, but a number of steps can be taken to control the amount and extent of this. Concrete structures designed and built in accordance with BS 8110[2] will have cracks naturally occurring at about 0.2 mm in width. These cracks do not affect structural performance but do affect permeability. If built in accordance with BS 8007[3] the concrete mix and reinforcement are designed to limit surface crack widths to 0.1 mm, thereby ensuring that the cracks will not extend across the complete cross-section of the wall and so reduce permeability. To achieve this, extra reinforcement is included within the concrete and a specially dense concrete mix is used. Great care is necessary in the design of expansion and movement joints, together with the design of reinforcement and concrete mix. Expansion and movement joints are detailed with 'waterbars' and 'stops' in an effort to render such joints watertight as shown in figure 7.23.

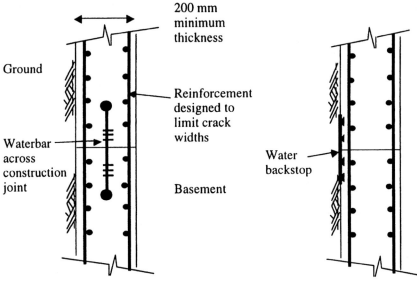

Figure 7.23

Such structures may in theory be able to stop the passage of water but cannot be vapour proof. Effectiveness of this type of construction relies greatly upon the standard of workmanship and detailing.

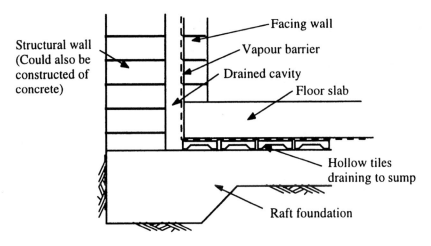

Figure 7.24

### 7.4.3 Drained Cavities

This system adopts the pragmatic approach and simply allows the ground water to penetrate the outer structural wall of the basement so that it may be dealt with inside. Once inside the wall, a false inner wall provides a cavity through which water can be drained away. Water vapour transmission is controlled by ventilation of the cavity and a vapour barrier can be applied to the inner wall to be doubly sure. A drainage cavity may also be provided under the floor, as shown in figure 7.24. This system may be used with waterproof concrete for those seeking maximum protection, but is not easily suited to multi-level basements as the intermediate floor construction must bridge the cavity to gain support. Of course, all of the above water protection methods can be used in combination to suit the Client's needs.

Table 7.3

| Grade | Use | Performance | Form of construction |
|---|---|---|---|
| 1 | Car parks and Plant rooms (excluding Electrical) | Some seepage and damp is tolerable | Type B Reinforced concrete only to BS 8110 |
| 2 | Workshops and plant rooms Retail storage areas | No water penetration but moisture vapour tolerable | Type A Type B to BS 8007 (watertight concrete) |
| 3 | Ventilated residential and working areas, Offices Restaurants and Leisure centres | Dry environment | Type A Type B to BS 8007 Type C to BS 8110 |
| 4 | Archives, stores and computer rooms | Totally controlled environment | Type A Type B to BS 8007 Combined with a vapour proof membrane Type C Ventilated wall cavity and vapour barrier to inner skin and floor protection |

*Extracts from BS 8102:1990 are reproduced with the permission of BSI. Complete copies can be obtained by post from BSI Customer Services, 389, Chiswick High Road, London W4 4AL*

### 7.4.4 Design Considerations

In choosing the right type of construction for the class of protection required, the Engineer must balance the initial cost of construction and subsequent cost of maintenance against the risk of water penetration. Such a choice will depend upon:

- The consequences of leakage.
- The feasibility of remedial and maintenance work.
- The risk of a changing ground water table levels, noting that BS 8102[1] recommends that we must consider ground water level rising three quarters of the way up the basement walls.
- The presence of aggressive ground conditions.
- The need to accommodate heave.
- Whether, in a confined site, access can be gained to the perimeter walls.

Table 7.3 is based on BS 8102[1] to give the reader a 'feel' for the type of construction needed for the class of protection required. The table relates grade of protection to the form of construction that is most likely to achieve it.

### 7.5 REFERENCES

1.  British Standards Institution. *BS 8102 Code of Practice for Protection of Structures Against Water from the Ground: 1990:* Complete copies can be obtained by post from BSI Customer Services, 389, Chiswick High Road, London W4 4AL
2.  British Standards Institution. *BS 8110 :Part 1: 1985: Structural Use of Concrete,* ISBN 0 580 14489 5
3.  British Standards Institution. *BS 8007: 1987: Code of Practice for the Design of Concrete Structures for Retaining Aqueous Liquids.*
4.  The Institution of Structural Engineers. *Design and Construction of Deep Basements: 1985*

# 8 Superstructures

The Client will initially appoint an Architect who will produce general layouts of the planned structure and approximate costs for various alternative forms of construction. On the approval by the Client of a particular scheme the Architect will appoint a Structural Engineer to carry out the detailed analysis of the structure. The precise time and brief for the Engineer will vary from structure to structure, but in most cases the Engineer will design all the structural members, and the foundations and prepare a series of structural drawings. These drawings will show the framing layout of structural members showing size of members and connection details. The structural drawings can then be used to communicate details of the structure to the steelwork fabricator or to other specialist suppliers.

The design of superstructures is very much the province of the Chartered Structural Engineer as it is considered to be a specialism within the industry. Many Structural Engineers are, however also Chartered Civil Engineers; both have similar skills and so the distinction is blurred. It is true that the design for any structure can become complicated and the Engineer must be familiar with all up to date British Standards, Codes of Practice and the commercial marketplace to produce a safe, economic design. The term superstructure is usually applied to construction of buildings above ground level as opposed to construction below ground, such as basement or foundations, which are termed the sub-structure.

Before the turn of the century the superstructure was generally constructed of brick, but this produced dark and dingy internal environments and a practical limit in height to four storeys due to the required thickness of load bearing brick walls. In 1880, soon after the development of the electric lift, an American Architect applied the frame idea to superstructure construction. The lift meant that multi-storey construction could extend upwards, unrestricted by the limitations of climbing stairs, and the open frame construction allowed buildings to be lighter in weight and have a lighter appearance inside by the ability to include a much greater window area.

The conventional superstructure is now a framed construction using beams and columns to support roof, floor and cladding. Competition between steel and concrete frames is fierce, and the popularity of each depends on small variations in price. Table 8.1 compares the variation of market share of different types of construction for buildings over two storeys during recent years. As can be seen from Table 8.1 there has been a small decline in steel frames recently, mainly due to the trend towards low rise buildings using load bearing brick.

The Client and Architect usually want as much uninterrupted floor space as

Table 8.1

|          | 1980  | 1992  | 1993  | 1994  |
|----------|-------|-------|-------|-------|
| Steel    | 33.3% | 56.9% | 61.6% | 57.8% |
| Concrete | 48.5% | 26.8% | 23.3% | 24.5% |
| Brick    | 14.3% | 13.7% | 11.9% | 15.6% |

*Figures courtesy of British Steel*

possible. This is because buildings are priced on floor area and must be flexible enough to accommodate varying internal partition arrangements as the use may change many times throughout their design life. Framed structures are good at providing open space floors, but they do have the disadvantage of internal columns which reduce internal space and dictate the layout of working areas. One step towards overcoming these disadvantages in medium rise structures is to use composite construction of steel and concrete to achieve long spans. Another popular solution is to use a central core, particularly in multi-storey structures. The floors are cantilevered from the central core which is used to house lift, stairs and service access. There are of course limits to the span of such cantilevered floors which limit the size of the building. Props at the end of the cantilevers provided by perimeter columns can greatly increase span but, these can affect the external appearance. The stability of structures against wind and accidental loading is important to the Engineer who must design the structures to withstand such forces. Stability can be provided by cross-bracing in steel frames, shear walls in concrete frames and secondary brick walls perpendicular to the main support walls in load bearing brick.

In this chapter we will take a look at three main categories of provision in the market:

- long span structures
- medium rise structures
- cladding to structures

Within these categories we shall look at the underlying engineering principles of the design, including stability and construction.

## 8.1 LONG SPAN STRUCTURES

It has long been the goal of the Engineer to increase the span of his structures to cut down costs and provide his Client with a clear unrestricted area for flexibility of use. Long span roofs are usually associated with low rise buildings of no more than one or two storeys in height, but they can be much higher, as in the case of aircraft hangers which stand on average 15 to 20 m height. Such structures can be constructed using portal frames, trusses, latticed portals, space frames and shell roofs.

### 8.1.1 Portal Frames

The idea of a portal frame came about as an optimised design of a simply supported beam. The strength of a simply supported beam is determined by the shape and the strength of the material of which it is made; such considerations give rise to an estimate of its 'moment resistance capacity' and can be readily calculated in accordance with today's design codes. 'Simply supported' means that the beam is supported on a theoretical knife-edge, free to rotate about the supports. The applied loading gives rise to a 'bending moment' which will cause the beam to sag and must not exceed the moment resistance capacity of the beam, although it is usually the deflection which is the limiting criterion. In the simply supported situation all the bending moment is concentrated in the centre of the span, as shown in figure 8.1. If we can spread the bending moment out and reduce the concentration at the centre, then the beam could either carry more load or increase its span. To achieve this we fix the supports so that they are *not* free to rotate. If the ends are considered fixed then the bending moment will be shared between the supports and the centre of the beam. Thus the beam is used more efficiently and forms part of a frame, called a 'portal frame', see figure 8.1.

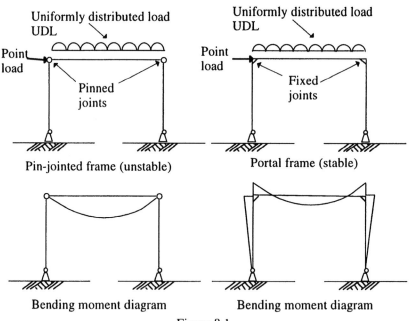

Figure 8.1

Figure 8.2 shows the general arrangement of a pitched portal frame, such frames being spaced generally at about six metres centres. Figures 8.5, 8.6 and 8.7 give details of typical connections. The pitched or sloping rafter behaves in a similar way to the flat-topped portal frame shown in figure 8.1. The joint

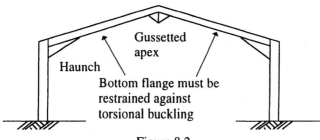

Figure 8.2

between the beams and columns is reinforced by a haunch, the purpose of which is to resist the bending moments attracted to this stiff joint. The gusset joint at the apex of the frame is also designed to resist bending moment attracted to the centre of the span. The advantage of a portal frame is that it provides a clear open volume for the Client, eliminating the need for struts and ties which are required for a trussed construction within the roof space. Aesthetics are also improved with the frame taking on a more 'modern' appearance. A portal frame can span distances of 12-45 m using commercially available steel section from British Steel, but it requires a complicated design with careful consideration being given to deflection and foundation movement. The size of the structural members is determined by Limit State Design to BS 5950[5]. Loading criteria is from BS 648[1] BS 6399[2] and CP 3: Chapter V[3]. There is an outward reaction, a spreading force, at the base of the frame due to loading and so precautions must be taken to prevent lateral foundation movement. This movement can be prevented, either by reliance on passive earth pressure on the outside of the base, or by using tie bars incorporated into ground beams between foundations.

*Variations on Design*

Portal frames are usually designed with pinned hinges at the base, as shown in figure 8.1. This means that the bottom of the column is free to rotate and therefore does not attract any moment. This simplifies the design and allows the Engineer to ignore the uncertainties of designing a moment resistive base. With pinned bases the stiffness of the structural members which make up the frame provide stability to the frame in the same plane as the frame. Fixed bases can reduce the size of the columns because the bending moments are spread out more evenly, but the foundations may have to be bigger to take the moments transmitted to the ground and so cost more.

Three-pin portal frames are also used. These reduce the bending moment in the roof beam and so its size may be reduced but, to control deflection, the column sizes will need to be increased. Advantages are a smaller roof void, simpler design and easier construction, but this design may increase the overall cost of the frame. Concrete portal frames are often constructed in this manner.

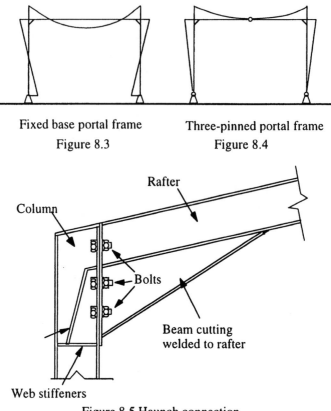

Fixed base portal frame        Three-pinned portal frame
Figure 8.3                          Figure 8.4

Figure 8.5 Haunch connection

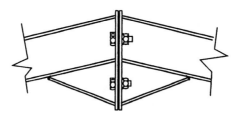

Figure 8.6 Rafter connection

### 8.1.2 Stability

The Engineer must consider the stability of a structure in all directions. This is often done by systematically considering each joint in a framed structure and considering a force applied in the x, y and z directions (north, east and up). The Engineer will look for mechanisms in the frame that need to be stabilised by adding bracing of some form. Stability of the structure in the plane of the portal frame is ensured by the stiffness of the frame itself and so lateral loads must be included in the design of the portal. Longitudinal stability, in the x direction, can be achieved by the use of 'X' braces, 'K' braces, or even secondary portable frames, as shown in figure 8.8, and 'W' bracing in the roof, as shown in figure 8.9.

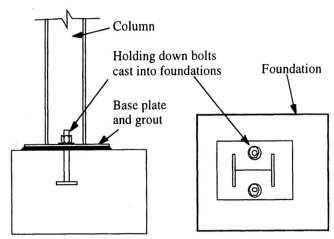

Figure 8.7 Base-plate details

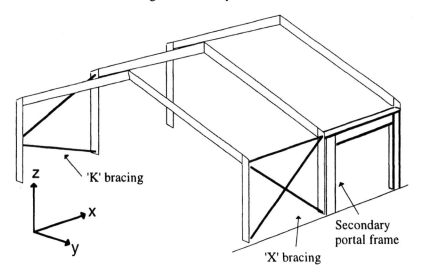

Figure 8.8

X bracing is the most efficient of the three. The principle of operation is to convert the force in the x direction to up and down forces within the columns. Keeping the bracing itself in tension, this keeps steel sizes to a minimum typically $50 \times 10$ mm flats. K bracing is a little less efficient, in terms of the weight of steel required to construct it, because the bracing members are required to carry both tension and compression loads. Such an arrangement may be necessary to accommodate an opening in the cladding, for example for a window. Secondary portal frames are the least efficient means of stabilising a structure and will only be used when the space in the frame is required for a large door opening and bracing cannot be accommodated elsewhere. Portal frames are not very good at controlling deflection and sections will need to be very stiff, making for an expensive form of stabilisation. The plane of the roof must also be braced in the x direction and this is often achieved by a W arrangement of bracing, as shown in figure 8.9.

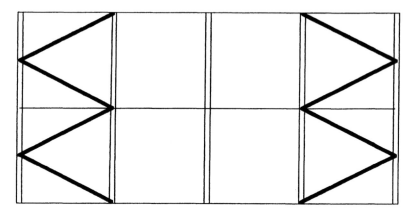

Figure 8.9 Plan view of roof bracing

Stability against uplift, in the z direction, is very important because suction forces can be induced on the roof in high winds. Uplift stability is ensured by holding down bolts cast into the foundations. The foundations are designed to be heavy enough to hold the building down in high winds.

### 8.1.3 Trusses

Truss construction is often used instead of a portal frame because it has a big advantage of being able to span large distances with relatively small magnitudes of deflection. Trusses are able to carry a huge variety of loading patterns which make them ideal for industrial applications. They are, however, difficult to fabricate and transport and can cost 20 to 30 per cent more than a portal frame. For trusses at three to six metre centres a rule of thumb guide to the proportions of a truss is that for spans of up to 30 m the depth of the truss would be about one fifteenth of the span, i.e. two metre deep and for spans greater than this up to say 90 m, a depth one tenth of the span would be necessary.

Two main types of truss are used today, a pitched and a flat top truss. Figure 8.10 shows the main types, plus an example of a bow string truss which is not often used nowadays because of the difficulty of fabrication. The slope on a pitched truss is usually determined by the limitation of the cladding to remain watertight and must be 1:15 or steeper. Pitched trusses are most commonly used in domestic construction as prefabricated timber trusses at 600 mm centres and can span up to 10 m.

A flat top truss is basically a lattice beam, usually constructed of steel angles or circular hollow sections and placed at six metre centres. The economic span limitation is about 45 m.

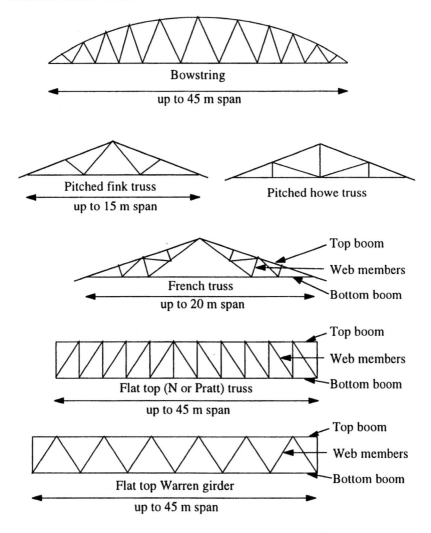

Figure 8.10

Apart from the advantage of a low deflection for span ratio the lower boom level can also be used to support a ceiling or loft floor. For larger trusses plant rooms can be incorporated within this roof void.

The disadvantages are that the trusses can take up a lot of internal volume of the building and the ties and struts obstruct services and can look unattractive. Loading must be applied at the joints of the frame and so it can also turn out expensive in terms of the extra steel required to carry a uniformly distributed load. The bottom boom also often requires extra bracing to stabilise it against compression in the event of load reversal from high suction forces on the roof due to wind loading.

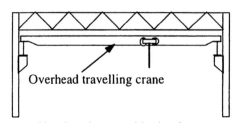

Simple column and lattice frame

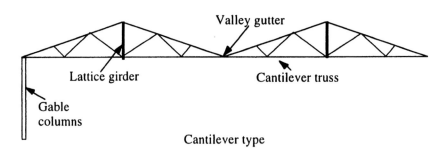

Cantilever type

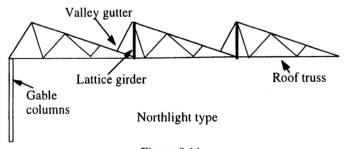

Northlight type

Figure 8.11

### 8.1.4 Latticed Portals

A typical lattice beam and column arrangement is shown in figure 8.11. Lattice beams and trusses can be used with fixed end conditions and are usually used when a flat roof is required. In a flat or low pitched portal the lattice beams have the advantage over portal frames of allowing the passage of services through the web of the beam. The bottom flange still requires restraint against load reversal due to wind uplift. Lattice construction is often used in conjunction with cantilever or roof trusses to allow natural light into the building. Figure 8.11 shows a few arrangements.

### 8.1.5 Space Frames

A space frame is basically a three-dimensional truss construction, all the joints between members are considered to be pinned and so carry no moment. The construction is shown in figure 8.12. Space frames are easy to transport, assemble and erect. Supplied in modules the frame can be assembled complete and then lifted into position or assembled in sections and then lifted into position. Space frames give a smart modern appearance allowing maximum light into the building.

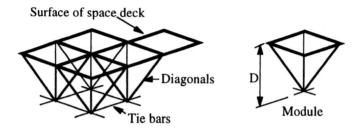

Figure 8.12

Such construction can span in two directions giving large clear unobstructed spans. Table 8.2 gives a guide to the spans possible with variations of depth. Of the Space Deck range the 1200 mm deep section is the most popular because it has stronger members and so can carry heavier loads.

Table 8.2

| Depth (mm) | Space Deck Ref. | Max. clear span (m) |
|---|---|---|
| 750 | 1275 | 23 × 23 |
| 1200 | 1212 | 36 × 36* |
| 1500 | 1515 | 40 × 40 |
| 2000 | 2020 | 50 × 50 |

* *This is the most popular size. (Information courtesy of Space Decks Ltd)*

Space frames can be more flexible than conventional construction and the entire building (including walls) may be constructed from space frame modules. Recent developments in space frame construction using moment resisting joints and advanced computer modelling have made possible frames that can span up to 130 m in each direction, but this type of design is expensive.

### 8.1.6 Shell Roofs

A shell roof may be defined as a structural curved skin whose strength is gained primarily from its shape. Usually constructed of concrete they are very costly due to the expense of the formwork. Shell roofs are also complicated to design usually requiring a finite element analysis on computer. Shapes include hemispherical and parabolic domes, vaults, saddles and shell shapes.

## 8.2 MEDIUM RISE STRUCTURES

Defined as up to four storeys in height, medium rise structures are the most common form of construction in the UK. Such structures are usually framed structures and can be constructed of concrete and/or steel. Load bearing brick construction is usually of the cross-wall type.

### 8.2.1 Concrete

Concrete frames have the advantage of being very flexible in form and provide the Client with a robust structure capable of carrying a wide variety of loading patterns should a change of use be required in the future. Concrete provides a structure which has a high degree of noise insulation from both without and within the structure. Concrete is however slower to construct than steel unless precast concrete members are used. Concrete structures can be provided either cast on site (in situ) or cast off site and transported to site for erection as precast units. Precast concrete is fast to erect and cheaper than in situ concrete. It is manufactured in factory conditions and so can be of high quality, but it does produce a relatively thick floor construction due to the down stand beams required to support the floors and can be as inflexible in design as steel. Concrete is expensive compared to other forms of construction and has had some highly publicised failures such as the Alkaline Aggregate Reaction (AAR); also if workmanship is poor then carbonation may be a problem.

### 8.2.2 In situ Concrete

An in situ concrete frame will consist of beams, columns and slabs cast on site between shutters with reinforcement placed inside. There are four types of floor construction: beam and slab, flat slab, waffle slab and a ribbed slab. Conventional beam and slab construction usually has columns and beams on a grid of 6 m centres with a concrete slab spanning between the beams, as shown in figure 8.13. Slabs can either span in one direction, as in figure 8.13(a), or in two directions, as shown in figure 8.13(b). Two-way spanning slabs can be slightly thinner than a one-way spanning slab, but they need more steel reinforcement and so can be more expensive. Table 8.3 shows the relative costs of the two forms of construction (1996 prices). The prices include the cost of steel and concrete only for slabs carrying normal office loading. In situ slabs are not less than 125 mm thick, because if thinner, it becomes difficult to include

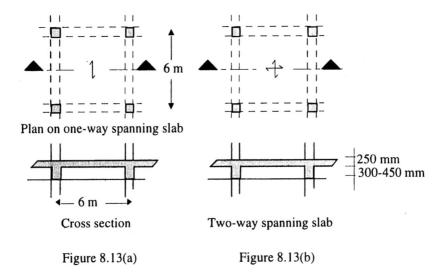

Plan on one-way spanning slab

Cross section

Two-way spanning slab

Figure 8.13(a)

Figure 8.13(b)

Table 8.3 Cost per m²

| Span and type | One-way spanning slab | Two-way spanning |
|---|---|---|
| 6 m beam and slab | £18.38 at 250 mm thick | £15.52 at 225 mm thick |
| 9 m beam and slab | £29.69 at 400 mm thick | £22.37 at 375 mm thick |
| 6 m flat slab | - | £46.80 at 350 mm thick |

Table 8.4 Span to depth ratios

| Span (m) | Beam and slab one-way spanning (mm) | Beam and slab two-way spanning (mm) | Precast beam and pot (mm) | Precast planks* (mm) | Flat slab (mm) | Waffle slab (mm) | Ribbed slab (mm) |
|---|---|---|---|---|---|---|---|
| 3 | 150 | 125 | 150 | 110 | 200 | - | - |
| 6 | 250 | 225 | 228 | 200 | 350 | - | - |
| 7.2 | 300 | 275 | 228 | 200 | 400 | 350 | - |
| 9 | 400 | 375 | - | 300 | 550 | 475 | 500 |

*Courtesy of Birchwood Concrete Products Ltd*

two layers of steel and fire regulations become a problem. Typical slab thickness to span ratios for different types of slab are shown in table 8.4.

The sizes shown in table 8.4 are a guide only, intended to give the reader a 'feel' for relative slab thickness and may vary depending upon individual circumstances. Slab and beam sizes should always be designed by a qualified Engineer for each particular case considered. These thicknesses assume a slab carrying a commonly taken office loading of 5 kN/m².

### 8.2.3 Flat Slab

The problem with conventional beam and slab construction is the thickness of construction. If we add the supporting beam depth to the slab depth the overall construction thickness of the floor can be 500 to 650 mm and if we then need to allow a void for services of say 750 mm the overall construction thickness of the

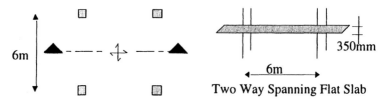

Figure 8.14

floor can be as much as 1.5 m. Large floor construction thicknesses can increase the overall height of the building and so increase cost. One way of reducing floor construction thickness is to use a flat slab. The principle behind a flat slab is that the beams are incorporated into the slab thickness so that no beam is visible on the underside of the construction. Such construction is often more expensive because it needs to use more steel reinforcement but can reduce overall building height and reduce shuttering costs.

### 8.2.4 Waffle and Ribbed Slabs

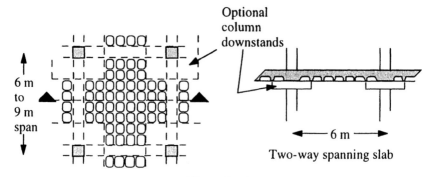

Figure 8.15

The intention behind the waffle slab is to use the advantages of the flat slab of thin construction but also to reduce the weight by casting voids into the underside of the slab, as shown in figure 8.15. Such construction is more efficient than the flat slab but is more expensive due to the complicated nature of the soffit formwork. It is usually only economic to use for spans greater than six metres and up to a maximum of nine metres. Both for flat slab and waffle slabs punching shear must be checked very carefully around the columns as this is usually the parameter that determines slab thickness. To help in this direction column downstands as shown in figure 8.15 are often used. Waffle slabs can span up to 14 m but these are not economic when compared to other forms of construction. Ribbed slabs are a one-way spanning version of the waffle slab. These are not often used today because of the more economic precast versions on the market.

### 8.2.5 Precast Slabs

Precast slabs are used in conjunction with the beam and slab system. The downstand beam can be of in situ concrete, precast concrete or steel. This type of floor is produced by specialist manufacturers and reference to their literature is recommended. Figures 8.16 and shows 8.17 the general arrangement of a plank floor and a beam and pot floor. Table 8.4 shows the span to thickness ratio for each type of construction assuming a 5 kN/m$^2$ live load.

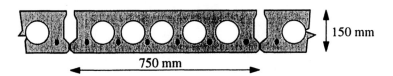

150 mm

750 mm

Figure 8.16 Precast plank floor *(Courtesy of Birchwood Concrete Products Ltd)*

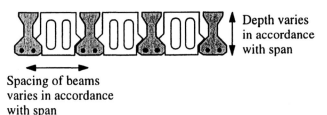

Depth varies in accordance with span

Spacing of beams varies in accordance with span

Figure 8.17 Beam and pot floor

### 8.2.6 Precast Concrete Frames

Precast concrete structural members have the same general dimensions as in situ concrete but are cast under factory conditions and transported to site for erection. The advantage of this system is that it is quick to construct and therefore cheaper. Structural elements are constructed under controlled conditions and therefore high quality units can be made and assembled on site. The disadvantages are a thick floor construction and it can be inflexible in design because all details of the structural dimensions have to defined before manufacture begins. This can be difficult to achieve in the early stages of a fast-track project. A typical precast beam and slab arrangement is shown in figure 8.18.

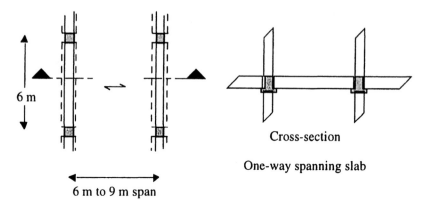

Cross-section

One-way spanning slab

6 m to 9 m span

Figure 8.18

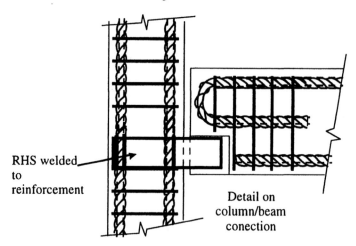

RHS welded to reinforcement

Detail on column/beam conection

Figure 8.19

Connections can be difficult with precast construction because all joints are assumed to be simply supported. Figure 8.19 shows a typical connection detail. Controlled conditions of manufacture, the possibility of using high strength concrete and prestressing tendons mean that the sizes remain about the same as can be achieved in in situ construction.

### 8.2.7 Stability of in situ and Precast Frames

This may be provided by reinforced concrete walls (shear walls) around stairs, lift shafts or as independent panels placed where window and door openings allow. Bracing panels can be constructed of high strength concrete block or brick but in that case care must be taken to consider differential shrinkage and creep between two materials. Of course, column/slab connections may be made moment-resisting in in situ designs, but this causes congestion of reinforcement. Stability of precast frames is achieved using shear walls, although cantilever columns may also be used spanning from the foundations.

Wind loads are resisted by the cladding and transferred to each floor level by spanning between floors. This load is then spread to shear walls via the 'diaphragm action' of the floor and is assumed, in a simplified design, to be resisted by the stiff walls in that direction.

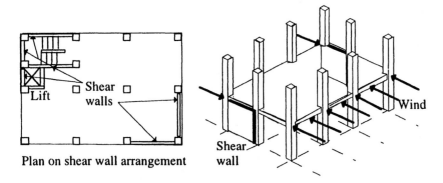

Plan on shear wall arrangement

Figure 8.20

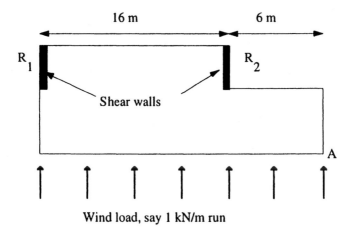

Wind load, say 1 kN/m run

Figure 8.21

The wind load is spread to each wall in proportion to its position and stiffness, the stiffer the wall the more load it attracts. As an example, we shall assume that each wall is the same stiffness but in non-symmetrical positions, as shown in figure 8.12. The reactions from each shear wall $R_1$ and $R_2$ are found by taking moments about $R_1$:

$$R_2 \times 16 = 22 \times 1 \times \frac{22}{2}$$

$$R_2 = \frac{22 \times 1 \times 22}{16 \times 2} = 15.1 \text{ kN}$$

Equating horizontal forces: $R_1 + R_2 = 22 \times 1$

$$R_1 = 22 - 15.1 = 6.9 \text{ kN}$$

Therefore     $R_1 = 6.9$ kN     and     $R_2 = 15.1$ kN.

Each shear wall can now be analysed for applied stress by using the following equations and the arrangement shown in figure 8.22.

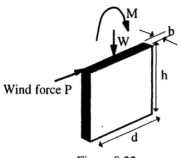

$$f = \frac{W}{b \times d} + \frac{M}{z}$$     where:

f = applied stress
b = wall thickness
d = wall length
h = wall height
W = load from upper stories or reaction from shear walls above
M = Applied moment = Ph = We.
(If the eccentricity e is greater than d/6 then tension will develop at one end of the wall.)

$$z = \text{section modulus} = \frac{b \times d^2}{6}$$

Figure 8.22

### 8.2.8 Steel Frames

The grid layout of a steel frame is very similar to in situ or precast concrete frames when using precast flooring. This is because the span capacity of the slab construction is the limiting factor. When using composite construction, however, larger spans may be achieved because the strength of the concrete floor and the steel beams are combined economically. Figure 8.23 shows the general layout of a typical steel frame using precast concrete flooring. The drawing provides beam sizes and reaction loads to the steel fabricator so that he may then be able to design the connection details. The precast units may be supported either on top of the beams or on angles fixed to the web of the beam. The problem with top seating of units is the resulting thick floor construction, but this can be overcome to some extent by providing holes for services within the web of the beam. Specialist fabrication of standard steel sections can be used to produce ready-made openings in the webs such as the castellated beam shown in figure 8.24. A popular solution to overcome thick floor construction is to seat the precast units in the webs of the beam, but this requires extra steel in the form of shelf angles and can significantly increase the weight of steel required overall. A screed of 50 to 75 mm thickness is applied to the top of the precast units to help bind the floor together so that it may act as a diaphragm and regulate the finished level.

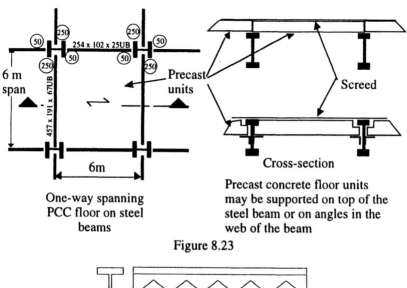

Precast concrete floor units may be supported on top of the steel beam or on angles in the web of the beam

One-way spanning PCC floor on steel beams

Cross-section

Figure 8.23

Figure 8.24

## 8.2.9 Composite Construction

This is a very popular form of construction due mainly to its speed and is probably the reason for the resurgence of the steel market share in the 1980s. The form of construction is shown in figures 8.25 and 8.26. Beams are placed at three metre centres with profiled metal sheeting spanning between them. This is used as a permanent shutter on which is cast the concrete topping. A shear connection is made between the concrete and steel by 16 mm diameter studs arc-welded through the deck onto the beam at 300 mm centres. This effectively welds the sheeting to the steel beams and when the concrete is cast the result is an efficient structural member using the concrete compression zone and the steel beam in the tension zone. The strength of the section can be varied by varying the size of beam and the thickness of the concrete topping. Stud centres can also vary depending on loading pattern. Table 8.5 shows a comparison of beam sizes required for both conventional and composite construction. Once again this information is provided to give the reader a feel for the size of beams required and may vary depending upon individual circumstances. Slab and beam sizes should always be designed by a qualified Engineer for each particular case considered. These sizes assume a commonly taken office loading of 5 kN/m² for the composite construction a concrete topping of 125 mm thick minimum and 150 mm thick precast units for the non-composite construction.

Table 8.5

| Span* (m) | Spacing (m) | Composite size *** (mm × mm × kg/m) | Non composite size ** (mm × mm × kg/m) |
|---|---|---|---|
| 6 | 3 | 254 × 146 × 31 UB | 356 × 171 × 51 UB |
| 7.5 | 3 | 356 × 127 × 39 UB | 406 × 178 × 67 UB |
| 9 | 3 | 356 × 171 × 57 UB | 533 × 210 × 82 UB |
| 12 | 3 | 457 × 191 × 82 UB | 610 × 229 × 125 UB |
| 15 | 3 | 610 × 229 × 125 UB | 762 × 267 × 173 UB |

*live load of 5 kN/m² is assumed **150 mm thick precast units on non composite beams
***125 mm thick minimum concrete topping on composite beams

As can be seen from the table there is a saving of approximately 40% in terms of steel weight and some advantage in smaller beam depths thus reducing the floor construction thickness. The most significant advantage of composite construction is, however, a rapid construction time known as 'fast track' construction. Although temporary bracing and propping may be required during construction of the concrete floors, the entire steel frame including stairs and decking can be erected in days. Steel is also flexible and versatile and can be welded on site should there be any last minute design changes.

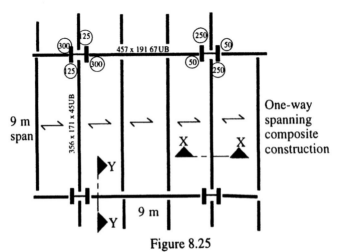

Figure 8.25

Stability of steel frames is achieved by the use of X-bracing or K-bracing, as shown in figure 8.8 for single storey, long span structures.

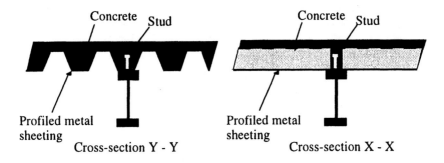

Figure 8.26

## 8.2.10 Load Bearing Brickwork/Blockwork

If the planned structure is medium rise and is highly compartmentalised, such as domestic flats or a hotel, then load bearing brickwork may be the most cost-effective structural solution. The layout of the walls must be such that they can support precast concrete floors of no more than six metres span; this is called a 'cross-wall' construction and consists of secondary brick walls perpendicular to the main support walls. The thickness of the internal load bearing wall should not be less than one 15th of the height and provide the precast units with

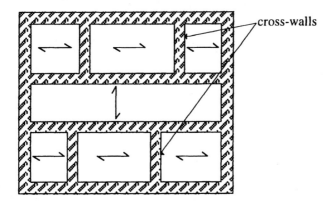

Figure 8.27

a seating of not less than 100 mm. Such walls provide stability for the structure as well as supporting the floor and upper storeys. In design the stresses in the walls will need to be checked to ensure that they do not exceed the capacity of the bricks or blocks used. Internal walls are usually constructed of blockwork, with 3.5 N/mm² strength required on the top two floors and 7 N/mm² blocks required on the lower two storeys of a four-storey structure. External cavity walls are constructed of 105 mm of brick, a 50 or 75 mm cavity and at least a 100 mm thick block wall. It is the internal block wall that carries the load of the building, the external leaf merely increasing the stiffness of the wall. Brickwork is subject to thermal movement and requires vertical expansion joints at 12 m centres. Blockwork is subject to shrinkage movement and requires vertical expansion joints at six metre centres. Horizontal expansion joints are needed every two storeys after the first three storeys. These are usually accomplished by provision of stainless steel angles fixed into the concrete at floor levels. A diagram of the support detail is shown in figure 8.28. External brick walls on the upper floors need to be checked to see if they can support the higher wind loads. Brickwork usually spans vertically, but can be designed to span in two directions as a panel if needed. It is not unusual for some form of concrete support or even a concrete inner leaf to be necessary to support wind loads on brickwork in high wind conditions.

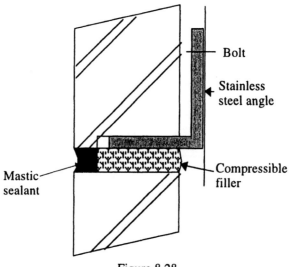

Figure 8.28

## 8.3 CLADDING

This refers to the material used to cover the sides of the building. There are five different forms of cladding:

- precast concrete
- steel sheeting
- curtain walling
- brick/block
- timber

The main purpose of cladding in general is to

- be self-supporting between the framing members
- provide necessary resistance to rain penetration
- be capable of resisting both positive and negative wind pressures and thermal movement
- provide resistance to wind penetration
- give required thermal insulation
- give required sound insulation
- give required fire resistance
- allow admittance of daylight and ventilation
- be easy to handle and erect
- give an attractive appearance.

### 8.3.1 Precast Concrete Panels

These are a common means of cladding and can be designed to fulfil the above specification. Precast concrete panels are often used in medium to high rise structures because of the comprehensive environmental protection they provide, but they are relatively expensive to construct and erect. In its simplest form the cladding arrangement is as shown in figure 8.29.

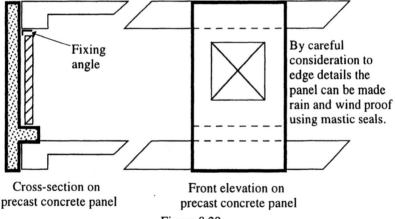

Cross-section on precast concrete panel

Front elevation on precast concrete panel

Figure 8.29

Considering the specification. precast concrete units span vertically by provision of reinforcement within the panel. It is recommended that only stainless steel reinforcement is used in this situation because of the highly aggressive environment and the potentially high cost of carbonation repair when in service. Carbonation is a result of the ingress of carbon dioxide and water into the concrete which neutralises the alkaline environment around the steel and allows it to rust. The rust pushes off bits of concrete, called 'spalling' and causes brown streaks on the outside of the concrete.

Reinforcement resists all wind pressures and the length of the panel is kept small so that thermal movement in minimised. Flexible joints are provided around the perimeter of the panels and slotted holes are provided in the supporting angles to allow expansion. Wind and rain ingress is prevented with mastic seals around the edges. Thermal insulation is provided by the inner block leaf and cavity insulation. Sound insulation relates to density of the material of the panel and with concrete this is high and so sound insulation is good. Fire resistance is naturally high with concrete. Windows can be easily cast in to provide flexible opening arrangements and the surface of the concrete can be coated with a vast range of finishes and colour to suit aesthetics. Panels are kept small to aid handling but most panels must be lifted by crane into position.

### 8.3.2 Steel Cladding

Profiled metal sheeting is often used as a low cost cladding material. It is quick and cheap to erect, but it does not provide complete protection from external environment such as noise and must be augmented with a block inner skin for working areas such as offices. It is usually only used on low rise construction and the general arrangement, in its simplest form, is shown in figure 8.30.

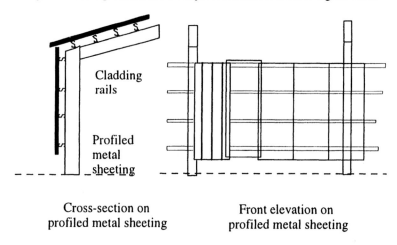

Cross-section on                Front elevation on
profiled metal sheeting         profiled metal sheeting

Figure 8.30

Profiled metal sheeting is supplied in sheets up to six metres long by 1200 mm wide and is fixed with self-tapping screws to cladding rails which span between columns at about six metre centres. These rails are cold-rolled sections and are designed to resist wind pressures and support the self-weight of the panels. The cladding panels themselves span 900 and 1200 mm between cladding rails. The thickness of the panels depends upon design and manufacturer and the external surface is often coated in a plastic film to prevent rusting and may also have an integral insulation lining.

Resistance to wind and rain is provided by overlaps of the panels horizontally and vertically. Thermal movement is not a problem in the panels themselves because the size is limited, but cladding rails are connected to the steel frame with slotted holes to allow movement to take place. Metal cladding will not provide adequate protection against sound, wind, heat loss or fire resistance, on its own, and will require additional measures internally, such as the provision of a block wall and insulation. Metal cladding has a wide range of finishes and colours to blend into surrounding architecture and can give a modern functional appearance. Windows and door openings can be provided easily by trimming the cladding rails around them.

### 8.3.3 Curtain Walling

Glazed curtain walling is the most expensive form of cladding and is able to give the building a modern stylish appearance. It consists of glazed panels on the full elevation of the building supported by mullions which span between floors. The mullions carry the full self-weight of the glazing and transoms and are designed to resist wind pressures. The general arrangement is as shown in figure 8.31.

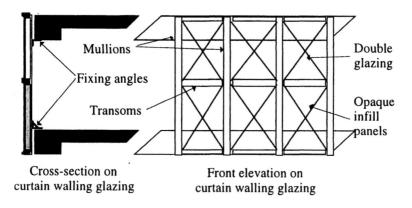

Cross-section on
curtain walling glazing

Front elevation on
curtain walling glazing

Figure 8.31

For self-support the mullions are fixed to the structural frame via metal angles which are provided with slotted holes to allow for thermal movement. Rain and wind penetration is resisted by mastic seals around the panels. Double glazing provides adequate thermal insulation, and some panels may be opaque and have a thermal insulation layer built within them. Sound insulation is good but can be improved by the provision of heavier opaque panels, an internal block wall and the use of triple glazing. Fire resistance is provided by 'intumescent glass panels' which expand to ten times their original thickness when exposed to heat, thus providing insulation. Obviously, this type of system provides many opportunities for window and door openings. The component parts are easy to handle and install, but a crane is often needed.

### 8.3.4 Brick and Block

Brickwork has long been a cladding for low and medium rise buildings. It is slow to erect, but is very durable and provides all the sound, wind and heat protection required for living quarters. It can be used as a cladding to a concrete or steel frame as shown in figure 8.32.

This type of cladding requires a compressible joint at the top of the blockwork to allow for differential creep. The Brickwork also requires a compressible joint every second storey to allow for differential Thermal movement.

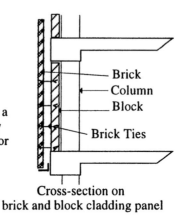

Cross-section on
brick and block cladding panel

Figure 8.32

Brickwork has its own flexural strength and acts compositely with the blockwork via brick ties to span either vertically or as a panel. Wind resistance is obvious but rain may penetrate the outer brick layer only to evaporate in the cavity. Consequently the cavity must be thoroughly ventilated. Thermal movements can be a problem and must be designed for with the inclusion of expansion joints. These are required vertically at 12 m centres and horizontally at 9 metre centres or at every third storey whichever is less. The horizontal joint may, however, be omitted if the building does not exceed 12 m in height, or 4 storeys. Details of a horizontal thermal joint are shown in figure 8.28. Thermal insulation is provided by the inner block leaf and cavity insulation and, because the construction is heavy, it gives good sound insulation. Brick is good for fire resistance and can be used to provide protection to the structure. Brick is a popular traditional building material which has a 'natural' feel and window and door openings can easily be provided.

### 8.3.5 Timber Cladding

This is normally only used for sheds and outbuildings, but metal veneered plywood is sometimes used for a mansard roof cladding as an alternative to roof tiles or slates. Timber studs or rafters span vertically; timber boards span between studs.

Timber weather-boarding provides some protection to rain and wind by overlap. Metal veneered cladding is system-designed to fit together and seal at the joints. Thermal expansion is usually not a problem for weather-boarding, but it can be for metal veneered boards and must be designed for at the joints. There is no thermal, sound or fire insulation, but windows can easily be provided.

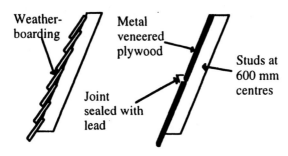

Figure 8.33

## 8.4 REFERENCES

1.  British Standards Institute. *BS 648: Schedule of Weights for Building Materials*, London 1964.
2.  British Standards Institute. *BS 6399: Design Loading for Buildings, Part 1: Code of Practice for Dead and Imposed loads,* London 1984.
3.  British Standards Institute. *CP 3: Chapter V: Loading, Part 2: Wind Loads*, London 1972.
4.  British Standards Institute. *BS 5950: Part 1 to 9: 1987 to 1995, Structural use of Steelwork in Building.*
5.  British Standards Institute. *BS 8110 :Part 1: 1985: Structural Use of Concrete,* ISBN 0 580 14489 5
6.  British Standards Institute. *BS 5268: Part 2: 1994: Structural Use of Timber,* ISBN 0 580 13790 2

# 9 Drainage

Drainage can be roughly divided into two categories: surface water drainage, which is designed to dispose of rain and excess ground water, and foul water drainage designed to dispose of sewage. The design of each is based on the same hydraulic principles. In this chapter we shall concentrate on surface water drainage, indicating the differences between foul and surface water when appropriate. The chapter begins by considering a simplified method of design and goes on to look at construction and some recent developments in trenchless technology. This chapter must be considered as an introduction to drainage design; more detailed information is available from the references made within the text, from the manufacturers and from the Water Research Council.

The strength of a road pavement depends on the strength of the subgrade (the ground upon which it is constructed) which, in turn, depends on the drainage system. The failure mechanism is simple; if water is allowed to seep into the subgrade it will eventually soften the soil to such an extent that the road pavement will fail. The drainage system is therefore designed to 'protect' the subgrade by controlling the amount of water in the soil and transporting it away from the road. The simplest and cheapest way to achieve water control of this kind is by the installation of simple gravity drains designed with enough capacity to cope with the removal of ground water, surface water and the occasional heavy storm.

## 9.1 DRAINAGE DESIGN

Drainage design is an iterative process where the Engineer has to make a number of assumptions and test them out via calculation. The method shown here follows the Wallingford procedure[9]. When involved in drainage design, we need to consider three aspects:

- Hydrological study            -rainfall quantity (how much rain?)
- Rainfall disposal system     -hydraulic design (design of the drain itself)
- The structural design of the drain to resist the weight of the soil above

### 9.1.1 Hydrological Study

A hydrological study is the study of water in and on the area of ground under consideration, which is called the 'catchment area', and is shown in figure 9.1.

For surface water drainage design we can limit our interest to rainfall quantity and we shall assume that our catchment area does not flood in the winter. We are only interested in high intensity, short duration storms to give a peak rate of 'run off' of water from the catchment area. To help us to estimate this, the Meteorological Office provides potential rainfall quantities for Great Britain based upon records kept over the last 100 years. A typical rainfall chart is shown in table 9.1 and gives the amount of rain which is likely to fall over a chosen period of time (called the return period). For example a storm of 2.5 minutes' length is likely to deliver 80.9 mm/h of rain once every two years or 101.8 mm/h once every five years. When designing surface water drainage for a road it is usual to consider a return period of two years. We can design for a return period of 5, 10 or even 100 years and in each case we would get larger drain sizes less likely to operate at full capacity. This can be considered as over-designed but it is up to the Engineer to decide what will be acceptable to his Client. The choice is between either an increased risk of occasional flooding due to an inadequate drainage system or higher construction costs resulting from over-design.

For highway drainage, our aim is to calculate the maximum flow of water which can be expected in the drainage system for a two-year return period. The maximum flow in a pipe at any point in a drainage system occurs when the entire catchment area is contributing to that flow. So the minimum length of time to consider for a storm is the time taken for a single drop of rain to fall on the farthest point of the catchment area and travel to the part of the drain that we want to design. The time taken for the rain drop to make that journey is called the 'time of concentration'.

Time of concentration = entry time + time of flow.

Thus the storm duration to consider is the time of concentration and for design purposes this depends upon the size of the catchment area under consideration and the length of drainage.

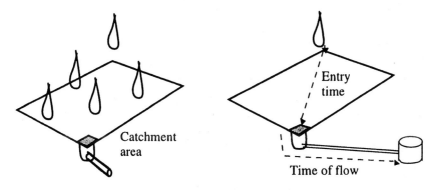

Figure 9.1                                   Figure 9.2

*Civil Engineering Construction*

Table 9.1 Typical Rainfall Intensity (mm/h) in Great Britain
*(Will vary with location.)*

| Duration | Return Period in years | | |
|---|---|---|---|
| (mins) | 1 | 2 | 5 |
| 2 | 69.2 | 85.8 | 107.8 |
| 2.5 | 65.2 | 80.9 | 101.8 |
| 3 | 61.6 | 76.5 | 96.5 |
| 3.5 | 58.5 | 72.7 | 91.9 |
| 4 | 55.6 | 69.3 | 87.9 |
| 4.5 | 53.1 | 66.2 | 84.3 |
| 5 | 50.8 | 63.5 | 81.0 |
| 5.5 | 48.7 | 61.0 | 78.1 |
| 6 | 46.8 | 58.7 | 75.4 |
| 7 | 43.4 | 54.7 | 70.6 |
| 8 | 40.5 | 51.3 | 66.5 |
| 9 | 38.0 | 48.3 | 63.0 |
| 10 | 35.9 | 45.7 | 59.8 |

*Crown copyright is reproduced with the permission of the Controller of HMSO
and Transport Road Research Laboratory*

Table 9.2 Drained Area of highway (m²) for R.I. of 50 mm/h and Channel flow
width

| Cross | | Gratings (1 m flow) | | Gratings(0.75 m flow) | | Gratings(0.5 m flow) | |
|---|---|---|---|---|---|---|---|
| fall | Gradient | HD/MD | kerb inlet | HD/MD | kerb inlet | HD/MD | kerb inlet |
| 1/60 | 1/300 | 128 | 90 | 65 | 55 | 22 | 22 |
| | 1/150 | 171 | 121 | 90 | 69 | 31 | 29 |
| | 1/100 | 205 | 149 | 108 | 84 | 38 | 35 |
| | 1/80 | 223 | 163 | 115 | 92 | 43 | 38 |
| | 1/60 | 261 | 182 | 137 | 107 | 50 | 45 |
| | 1/40 | 270 | 193 | 163 | 120 | 59 | 54 |
| 1/40 | 1/300 | 250 | 201 | 128 | 107 | 43 | 43 |
| | 1/150 | 338 | 260 | 177 | 143 | 61 | 58 |
| | 1/100 | 386 | 324 | 212 | 177 | 75 | 70 |
| | 1/80 | 431 | 335 | 230 | 198 | 84 | 76 |
| | 1/60 | 462 | 363 | 265 | 231 | 96 | 90 |
| | 1/40 | 566 | 444 | 325 | 248 | 116 | 110 |

*HD/MD heavy duty/medium duty gully grating*
*Information courtesy of Transport Road Research Laboratory*

The entry time is the time taken to enter the drainage system from the most remote point of the catchment area and will vary with gradient. By using the catchment areas of highway, as shown in table 9.2, and as recommended in TRRL Contractors Report 2[12], we can simplify the entry times to three minutes for highways and five minutes for car parks, irrespective of the distance across the catchment area to the first gully. Time of flow is the time taken for the water to travel the length of the drainage system 'upstream' of the point under consideration. This time can only be obtained by assuming a size and gradient of drainage pipe to be used and then reading off the hydraulic velocity from the hydraulic graph shown in figure 9.3. The distance divided by the velocity gives the time of flow. Establishing the time of concentration allows the rainfall intensity to be read from table 9.1 and then the maximum quantity (or flow) of water which can be expected to be carried by the drain can be calculated from the following equation.

$$Q = 3.610 \text{ A I p (litres/sec)}$$

This is called the Modified Rational Method (or the Lloyd Davies formula)

Where:  Q = Runoff Quantity (litres/sec)
A = catchment area (ha) (1 ha = 10,000 m²)
I = rainfall intensity (mm/h)
p = proportion of runoff after allowing for soakage and evaporation and depends on permeability of the surface and soil (usually 1.0 for highways/car parks and 0.6 for grass)

## 9.1.2 Hydraulic Design

Once we have calculated the maximum flow of water for our drainage system we then need to design the size of pipe required to deal with that flow. Here we shall look at airtight pipes since this will apply to most surface (and foul drains).

The flow (or average velocity) of water in a pipe depends on gradient, drag and size. The steeper the gradient, the faster the flow and together with the size of pipe are the parameters we can control in the design to adjust the flow. The variation of these parameters is given in the hydraulic flow chart shown in figure 9.3. (A different flow chart is required for foul water design.)

Drag is the name given to the friction between the fluid and the side of the pipe and is a function of the texture of the walls of the pipe and the suspended solids in the fluid, i.e. the smoother the walls, the faster the flow. The amount of drag is represented by the symbol '$K_S$'.

$$K_S = 0.6 \text{ mm for surface water}$$
$$K_S = 1.5 \text{ mm for sewage}$$

The $K_S$ number will vary depending upon the material from which the pipe is made and specialist manufacturers' information should be consulted for pipe performance.

The 'flow' of a pipe also depends on how full it is and there are two reasons for this. First, the greater the depth of the water in the pipe then the greater is the area of pipe in contact with the liquid and so the greater is the drag, and second, the flow is faster if the flow is laminar rather than turbulent. The term 'laminar' means a smooth, streamline flow with no disturbance; the flow is turbulent if the flow is disturbed. If turbulence occurs then the average flow velocity decreases and blockages are more likely. It can be shown that a pipe is most efficient operating at about 80 % capacity and that is what the Engineer will aim for in the design.

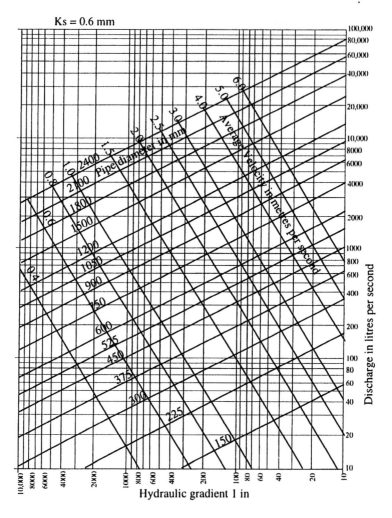

Figure 9.3 Hydraulic flow chart
*This chart is based on information provided courtesy of ARC Pipes[1]*

## Example 1

Consider a gully draining an area of car park 880 m², with a time of entry of five minutes and return period of two years. The distance from the inlet gully to the first manhole is 30 m and this is the pipe we wish to design.

First select a pipe and gradient, then we see if it will be sufficient from the calculations. Choose (say) 150 mm diameter at a gradient of 1 in 100 (say). Then from the hydraulic flow charts, in figure 9.3, we can see that the average velocity of flow is one metre per second.

$$\text{So the time of flow} = \frac{30}{1} = 30 \text{ seconds.}$$

Time of concentration is therefore $= 5 + 0.5 = 5.5$ minutes

When the time of concentration has been reached, the furthermost parts of the catchment area are all contributing to the flow in the pipe. The maximum flow has therefore been achieved at this storm duration.

From Table 9.1 rainfall intensity (2 years return) $= 61.0$ mm/h.

Run off quantity Q at the end of the first 30 m of pipe:

$$Q = 3.610 \times \frac{880}{10000 (\textit{turns } m^2 \textit{ into hectares})} \times 61.0 \times 1.0 = 19.4 \text{ litres/second}$$

This quantity now needs to be compared with the capacity of the chosen pipe, as shown on the hydraulic flow chart in figure 9.3.

From figure 9.3, for a 1:100 gradient and a 150 mm diameter pipe, maximum discharge, running full equals 18 litres/second. This is not enough and the 80 per cent capacity has been exceeded; to over come this we can either increase the gradient to 1:20 or increase the diameter to 225 mm.

For a network of drains the calculations are cumulative and can become complicated, so the design sheet shown in figure 9.5 is used.

*Example 2*

It is required that we size the pipes shown in the drainage system shown in figure 9.4. Entry time is five minutes (say) and return period is two years. Assume that the maximum gradient is limited to 1 in 200.

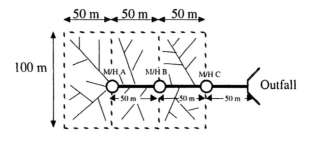

Figure 9.4

| Drainage Design Table | | | Site name and location | | | Job number | | | Date | | | |
|---|---|---|---|---|---|---|---|---|---|---|---|---|
| Entry time = 5 mins | | | Return period 2 years | | | Proportion of run off = 1.0 | | | | | | |
| Length and reference | Upper invert level | Lower Invert level | Fall | Distance | Grade | Pipe diameter | Max average velocity | Time of flow | Time of concentration | Rainfall intensity | Catchment area | Cumulative total area | Run-off quantity | Capacity from hydraulic flow chart | Proportion of capacity |
| (m) | (m) | (m) | (m) | (m) | (1 in -) | (mm) | (m/s) | (mins) | (mins) | (mm/hr) | (m²) | (hect) | (ltrs/sec) | (ltrs/sec) | |
| A-B 50 | | | | 50 | 200 | 150 | 0.7 | 1.2 | 6.2 | 58 | 5000 | 0.5 | 110 | 13 | ✗ |
| | | | | | | 300 | 1.1 | 0.8 | 5.8 | 60 | | 0.5 | 110 | 77 | ✗ |
| | | | | | | 375 | 1.3 | 0.6 | 5.6 | 61 | | 0.5 | 110 | 140 | ✓ |
| B-C 50 | | | | 50 | 200 | 525 | 1.6 | 0.5 | 6.1 | 59 | 5000 | 1.0 | 210 | 340 | ✓ |
| C-D 50 | | | | 50 | 200 | 525 | 1.6 | 0.5 | 6.5 | 56 | 5000 | 1.5 | 300 | 340 | 88% |
| | | | | | | 600 | 1.7 | 0.5 | 6.5 | 57 | | | 310 | 480 | ✓ |

*Note - all calculations are correct to two significant figures.*

Figure 9.5 *Drainage design chart*

(*Crown copyright; reproduced with the permission of the Controller of HMSO*)

The design calculations are shown in figure 9.5 presented in the Engineer's standard format as recommended by the Wallingford procedure[9]. The reader will note from this design sheet that the time of concentration will increase with each subsequent run of drainage.

## 9.2 STRUCTURAL DESIGN OF DRAINAGE

Once we have selected the diameter of our drain we now have to consider its structural strength. This depends upon:

- The strength of the material it is made of
- Bedding and surround detail
- Depth of drain and width of trench
- Surface loading

When considering the strength of the material from which the pipe is made, manufacturer's advice must be sought. For concrete pipes we use class L, M and H for low, medium and high strength concrete. For clay pipes we use S, ES and SS for Standard Strength, Extra Strength and Super Strength. The load which can be supported by a pipe is a function of the bed and surround detail, the strength of the pipe and the width of trench. Figure 9.6 shows typical surround details currently used by the Department of Transport; these will vary from the recommendation of the specialist manufacturer. The trench width is 300 mm wider than the external diameter of the pipe.

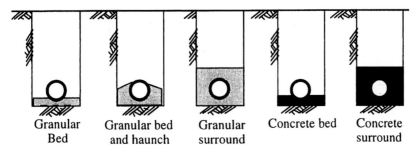

| Granular Bed | Granular bed and haunch | Granular surround | Concrete bed | Concrete surround |

Figure 9.6

The specification of bedding arrangements and the depth restrictions shown in figures 9.7 and 9.8 are specified by the Department of Transport's (DoT) Advice Note HA 40/89[8]. These tables relate to loading conditions consistent with a drain installed under a road and so account for the severest degree of expected loading. It is true to say that the DoT recommendations are more conservative than those recommended by the manufacturers and if the drain is to be designed for any other Client, then the manufacturer's own design manuals should be consulted.

### 9.2.1 Joints

All pipes must be jointed and typical pipe joints are shown in figure 9.9. The joint must be airtight and flexible and this is achieved with rubber seals incorporated within the joint. Airtight joints prevent leakage, whilst flexible

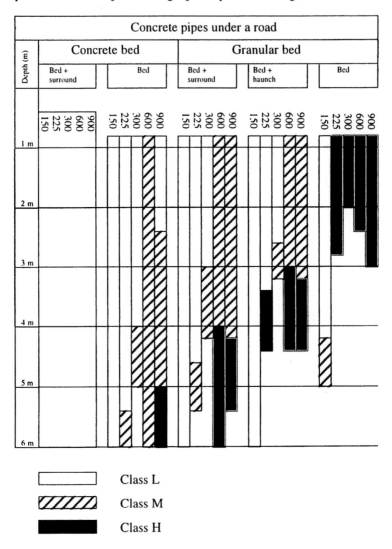

Figure 9.7
*Information extracted from DoT Advice Note 40/89[4]*
*Crown copyright; reproduced with the permission of the Controller of HMSO*

joints allow settlement movement to take place after the pipe is installed. This avoids over-stressing the joint and possible failure of the pipe. The concrete surround must also have a soft joint placed in the concrete at pipe joints to maintain the flexibility of the pipe run. Once constructed, all pipe runs must be tested for airtightness by undergoing an air pressure test or water pressure test in accordance with DoT guidelines, see section 9.3.3.

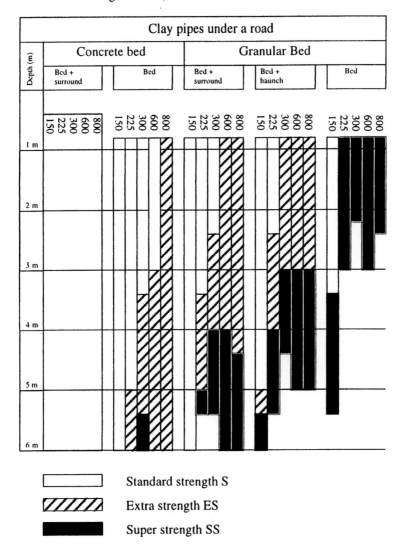

Standard strength S

Extra strength ES

Super strength SS

Figure 9.8
*Information extracted from DoT advice Note 40/89[4]*
*Crown copyright is reproduced with the permission of the Controller of HMSO*

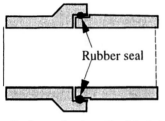
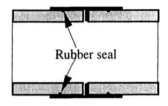

Socket and spigot flexible joint          Sleeve flexible joint

Figure 9.9

### 9.2.2 Sulphate Conditions

Exposure to sulphate attack can cause concrete to dissolve and crumble. It is, therefore, very important with concrete pipes to check the exposure to sulphate and hydrogen sulphide attack. These chemicals can occur naturally in the ground around the pipes, in ground water and in the water or sewage carried by the pipes. Sulphate-resisting cement can be specified in accordance with BS 8110[2] but this does not protect the drain from hydrogen sulphide attack. Clay pipes are immune from attack by both chemicals. Exposure to such an environment is particularly acute for foul water drainage and so clay pipes should be used wherever possible.

### 9.2.3 Vitrified Clay Pipes

These will have fewer conservative surround specifications, allowable depths and loadings than those specified by the DoT. For design information it is recommended that advice be sought from the Clay Pipe Development Association[7].

### 9.2.4 Manholes

Figure 9.10 shows a typical manhole construction using precast concrete rings. Again, there are restrictions on depth due to loading and reference must be made to manufacturer's design tables. Manholes are usually placed at a maximum of 80 m centres, and are located at changes in gradient, pipe diameter or changes in direction.

### 9.2.5 Gullies

Figure 9.11 shows a typical gully detail and table 9.2 gives one method of assessing the spacing of gullies. Width of flow of 0.5 m indicates the width of water flowing in the channel of the road adjacent to the kerb during a storm. Many roads have one-metre hard strips adjacent to the traffic lane on the carriageway, and it is prudent to allow this to take temporary run-off during storms in order to economise on gullies. The width of flow is determined at 0.5 m, 0.75 m or 1.0 m depending on the class of road and proximity of pedestrians.

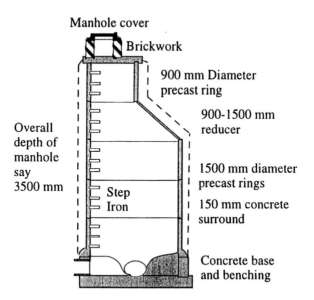

Figure 9.10

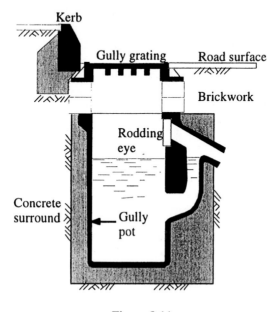

Figure 9.11

### 9.2.6 French Drains

French drains or filter drains are usually laid using a rule of thumb 30 m length of 150 mm diameter perforated pipe at a fall of 1 in 200 which is adequate for most situations for the first run of pipe. Thereafter, the diameter of the pipe, size of perforations and grading of stone surround must be designed from estimates of the anticipated flow. The typical general arrangement is shown in figure 9.12.

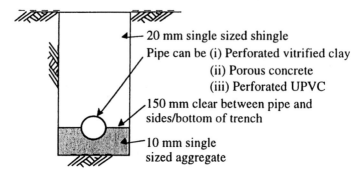

Figure 9.12

## 9.3 DRAINAGE CONSTRUCTION

Traditional methods of construction are to simply dig a trench to the correct depth and fall, lay the pipes by hand on gravel, 150 mm clear of the bottom, surrounding them with gravel by hand and then backfill by machine. The excavation of the trench must be carried out accurately and safely. Accuracy is achieved by profile boards or laser and safety is achieved with trench sheets. The width of the trench is set by the width of the bucket of the machine, so the right choice is essential.

Any excavation deeper than 1.2 m must be supported against collapse. If the water table is higher than 1.2 m then even shallower trenches may need support depending on the type of ground. Much of a Civil Engineer's work will be below ground and it is essential that sufficient thought is given to the safety aspects. Support is required for all excavations, because even in apparently stable ground the sides can collapse in seconds without warning. Soil need only come up to a man's waist for the pressure to be such as would break a leg. If the depth was any greater then the operative could be crushed. Support is achieved by timber shoring, propped steel sheet piles or trench shields or boxes. Figure 9.13 shows traditional timber shoring. Here 225 mm by 25 mm timber boards are pushed against the sides of the trench using the trench props (sometimes called winding props); these support the soil whilst drain laying is in progress. This type of support is time consuming to install and costly since the timber poling boards often get broken either during use or in handling. The inherent weakness of timber also limits the depth to which the material can be used safely to 2 to 2.5 m depth depending on soil type. It is often easier to install the support system along the entire length of drain under construction before pipe laying commences. This gives the advantage of being able to adjust the line and level, inspect and then test the entire length of drain before back filling, but obviously this is expensive and the method must be considered on its merits.

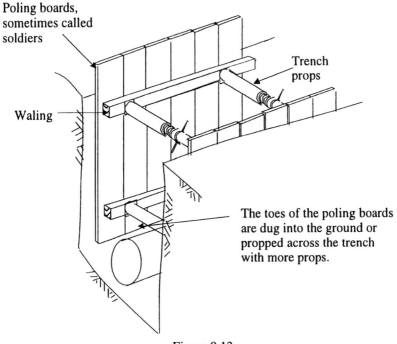

Poling boards, sometimes called soldiers

Trench props

Waling

The toes of the poling boards are dug into the ground or propped across the trench with more props.

Figure 9.13

To overcome the problem of weakness of timber, steel trench sheets are often used up to 2.5 to 3 m depth. A cross-section of the steel trench sheet commonly used is shown in figure 9.14, with the steel 2.64 mm thick.

40 mm

380 mm

Figure 9.14

These sections do not have clutches which interlock and therefore cannot be considered to be watertight. Propping is still required in the same way as for timber and the speed of installation is still slow. They do, however, provide the essential protection needed for men working down the trench. To overcome the problem of speed of installation, prefabricated trench shields or boxes are sometimes used. The arrangement is shown in figure 9.15. Only two pairs of these shields are usually required and are used in a leapfrog process as the trench is dug protecting just the working area. This requires that each joint of the pipe is set at the correct line and level and partially surrounded before proceeding to install the next pipe. Since safe access to the trench is so brief there is inevitably greater risk carried by the Contractor in the event of a failed inspection or test, but with an experienced drainage gang the advantages of speed far outweigh these risks.

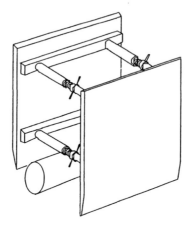

Figure 9.15

### 9.3.1  Line and Level

The correct line and level of the drain is conventionally defined by the erection of profile boards which is a job usually assigned to the site setting-out Engineer. The concept behind profile boards is to define the correct level and fall using two or three profile boards set at a constant distance (say 2 m) above the inverts of the drain pipe, as shown in figure 9.16. The profile boards are erected above ground so that a clear line of sight along the profiles can be seen along the trench. The line marked out by the profiles is then the gradient (or fall) of the pipe to be installed. The constant measurement down from the gradient is provided by a 'traveller' which is a length of timber, two metres long with a cross-board at the top. The top of the traveller is eyed into the gradient marked out by the profiles by looking along the top of the profiles and moving it up or down until the top of the traveller is in line with the profiles.

Increasingly, with pressure on time and money, laser guidance is used to achieve line and level accuracy. With this method a low-powered laser is set up at the start end of the trench so that the base of the laser is at the same invert level as the pipe being installed. The desired gradient of the drain is entered into the machine and a laser light beam is shone down the pipe at the correct fall. A laser target is placed in the pipe at each successive joint as construction progresses and if the pipe is in the right location a small red dot will appear in the centre of the target as shown in figure 9.17.

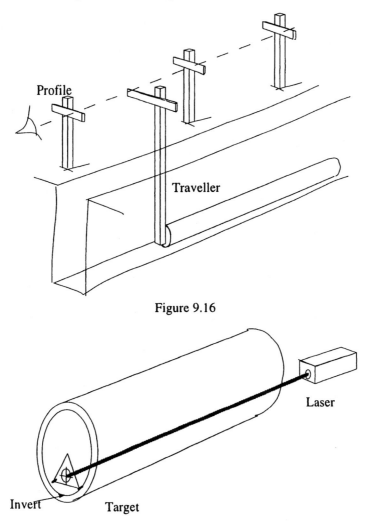

Figure 9.16

Figure 9.17

## 9.3.2 Backfilling

This operation must be carried out with care to avoid disturbing the pipes and to provide reliable compacted ground over the pipe which may carry a road construction. The backfill operation is usually carried out in two stages. The first stage is where the shingle bed and haunch (usually 10 mm single size shingle) is carefully pushed around the pipe and trodden in during the adjustment operation for line and level. When line and level are correct and the shingle surround is completed, the construction gang must ensure that there is a minimum of 150 mm thickness of shingle between the outside of the pipe and the side of the trench. The trench is then backfilled with suitable soil in layers of 150 to 200 mm thickness. Suitable soil is usually the soil excavated from the trench in the first place; this is called the 'arisings'. Under a new carriageway, however, backfilling is carried out with suitable imported granular material. Each layer is compacted either with a vibrating plate compactor ('a wacker plate'), an impact compactor, or a trench roller. Obviously, care must be taken when compacting the layers immediately above the pipe. All backfilling carried out in existing roads must now conform to the Highway Authorities and Utilities Committee (HAUC) Specification[10].

### 9.3.3 Testing

Testing must be carried out to ensure that the complete drainage system works correctly. For surface water drains, pipe runs do not have to be airtight, but they are expected to pass a dye test and a visual inspection. For the dye test, dye is introduced into the system upstream of the area under inspection. Dyed water is then followed visually through each manhole. A visual inspection simply requires that an observer looking down one end of a pipe can see the 'full moon' of the end of the pipe at the next manhole. For sealed drains (including foul drains) each pipe run must be tested for air tightness by inserting a plug at each end of the run and an increased air pressure measured by a manometer at one end. The increase in air pressure is achieved by blowing down a rubber tube attached to a bleed nipple on one of the plugs. The pressure is not great, but similar to blowing up a balloon. The bleed nipple is closed and a steady pressure must be maintained for a set period of time. An alternative to the air test is the water test; details for both these tests are given in BS 8301[3] and BS 8005[4].

### 9.3.4 Trenchers

These machines are sometimes used for the construction of long lengths of pipe ducts or cables. They are sometimes used for free gravity drain construction because the depth of dig can be continuously varied, but accuracy of the gradient of the excavation is difficult to achieve. The principle of operation is a conveyor of small buckets or shovels which cut into the ground rather like a chain saw cuts into wood. See figure 9.18.

### 9.3.5 Slot Trenching

A variation of the conventional method is that called 'slot trenching'. This method is used to install cables and ducts at shallow depths of up to 1.5 m. The trench is cut with a 1.5 m diameter disc 100 to 200 mm wide. The cable or duct is laid as the machine passes and the material excavated is pushed back into the hole and compacted in one operation. See figure 9.19.

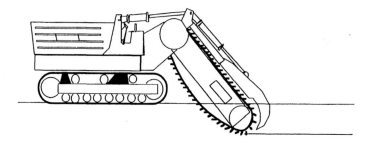

Figure 9.18

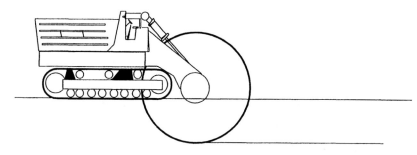

Figure 9.19

### 9.3.6  Roads and Street Works Act 1993

The Roads and Street Works Act 1993[11] was implemented with the aim of achieving fewer excavations leading to less disruption to the road user. The objectives were to improve co-ordination between utilities, thus reducing the number of holes dug in our roads and to improve the speed and quality of the reinstatement work by allowing the utilities to reinstate their own trenches instead of paying the Local Authority to do it. The Specification for the Reinstatement of Openings in Highways[10] was drawn up by the Highway Authorities and Utilities Committee (HAUC). The Specification gives a list of acceptable materials and methods for the reinstatement of holes in the roads and requires that the work be supervised by a specially trained member of staff. The Act has been successful in improving the speed and quality of the work, but it has had implementation problems. The Act has had only partial success since the co-ordination of such work is difficult. The work is supposed to be co-ordinated via a national register for services which all utilities would consult; the computer software is planned to be introduced in 1996. In the meantime the aim of the Act is being aided by the introduction of modern trenchless pipe and cable laying machines.

## 9.4 TRENCHLESS TECHNOLOGY

As we have seen there is a lot of effort involved in laying a drain in a
conventional trench. If the drain is to be constructed in an urban situation there
will be a great deal of disturbance and inconvenience to road users and the local
people. There are methods available which allow the construction of a drain
without digging a trench, but the potential of these has not yet been fully
exploited although they are likely to be more commonly used in the future.
There are three main methods available for the trenchless installation of drains,
pipes or ducts: these are thrust boring, directional drilling and pipe bursting.

### 9.4.1 Thrust Boring

Thrust boring is a method of jacking the pipe through the ground between the
launch pit and the receiving pit. The arrangement is shown in figure 9.20. The
pipe is headed by a small diameter boring machine which cuts its way through
the ground under laser and computer control. A rotating conical cutting head
with tungsten carbide cutters cuts the soil into small pieces as it is pushed along
by hydraulic jacks. The soil pieces are channelled back through the boring head
into a crusher and the resulting mixture is sucked to the surface with a vacuum
pump. As the machine moves along additional sections of pipe are added in at
the launch pit; insertion of a new pipe takes about 15 minutes. Diameters from
250 mm to 1.05 m can be accommodated and can be driven up to 200 m.

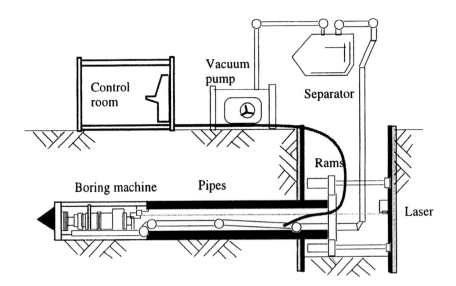

Figure 9.20
*Diagrams is based on information courtesy of The New Civil Engineer*

## 9.4.2 Pipe Bursting

If an existing pipe needs to be replaced, a system called 'pipe bursting' can be used, as shown in figure 9.21. In this situation a slightly larger pipe is pushed along the old pipe line using hydraulic rams. At the front of the new pipe is a metal cone which, as it is pushed into the old pipe, bursts it outwards, compacting the pieces into the surrounding fill to make way for the new pipe. The steel cone can also be drawn along by a steel wire but the effect is the same. This system has obvious advantages of speed and ease of installation, but if the existing pipe has branches along its length then these will be severed.

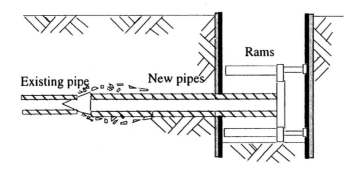

Figure 9.21
*Diagram is based on information courtesy of The New Civil Engineer*

## 9.4.3 Directional Drilling

Directional drilling is used to install ducts, cables and service pipes. It is used for crossing roads, rivers or weaving through existing services if their positions are known. The system consists of a cutting head pushed through the ground by a string of flexible steel tubes, known as the drill stem, from the surface machine as shown in figure 9.22. The cutting head is about 50 mm in diameter and it can be a mechanical cutting device something like a rotating drill bit, or a small diameter nozzle through which water is forced at high pressure, thus 'cutting' into the soil. The sequence of operation is that first the drill bit is started at the launch pit and drilled approximately horizontally through the ground to a pre-dug reception pit. The position, depth, tilt angle and the axial orientation of the cutting head are monitored by radio signals received by a portable monitor held by a skilled operative on the surface. Direction and depth of the cutting head can be altered continuously by radio contact back to the drill operator. The orientation and thrust can be altered from this position.

Once the drill head is in the reception pit, the drill head is removed and a cutting and compacting mole is attached to the drill stem, as shown in figure 9.23. The service pipe is then attached to the cutting and compacting mole. The drill stem is then pulled back through the ground to the surface machine pulling the cutting and compacting mole and the service pipe with it. In one operation the mole enlarges the hole from 50 to 200 mm diameter, compacts that soil into the surrounding ground and pulls through the service pipe.

The surface machine can push or pull with a force of up to 8 tonne which is sufficient to drill up to 200 m. The push force must be limited to prevent buckling of the drill stem; the pull force can be greater.

Radio link between 'guide man' and surface machine operator

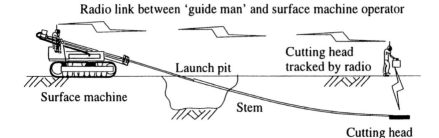

Figure 9.22

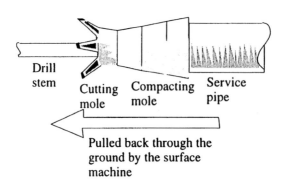

Figure 9.23

The most serious problem with trenchless technology is the avoidance of existing pipes and services. At present, utility advice to the Contractor is to approximately locate by radio or metal detectors the suspected position of any existing service and then to dig down by hand to establish its precise location and identity. This can be a difficult, time-consuming and expensive operation but the cost of hitting and disrupting a service can run into tens of thousands of pounds. A drilling or boring machine can easily cut through an existing service unintentionally and it is in this area that research and development is so desperately needed. There are basically two aspects to the problem; the first is that the precise location of existing services is difficult. The second problem is that drilling and boring machines cannot 'see' where they are going and will not know of an existing service until it is too late. The location of existing services is being tackled by an integrated national computerised register of service locations which is intended to be accurate, but since many locations are recorded plus or minus three metres, there is a lot of work to be done in this area before the system can become reliable. The detection of underground services from the drill head or the surface is an area that requires further research and development. Trenchless techniques for pipe and service installation and repair are already very useful, but hold the potential for great benefit to our society in the future if the problems of existing service location can be solved.

## 9.5 REFERENCES

1. ARC Pipes. *Design Manual,* 1990. Mells Road, Frome, Somerset BA11 3PD
2. British Standards Institution. *BS 8110 :Part 1: 1985: Structural Use of Concrete,* ISBN 0 580 14489 5
3. British Standards Institution. *BS 8301 British Standard Code of Practice for Building Drainage:* London 1985
4. British Standards Institution. *BS 8005: Parts 0,1,2, and 4: 1987. British Standard Code of practice for drainage (Part 3: 1989 and Part 5: 1990)*
5. British Standards Institution. *BS 6367: 1983 Drainage for Roofs and Paved Areas*
6. British Standards Institution. *BS 8301: 1985 Building Drainage*
7. Clay Pipe Development Association 30 Gordon Street, London WC1H OAN.
8. Department of Transport. *Advice Note HA 40/89: Determination of Pipe and Bedding Combinations for Drainage Works.* HMSO ISBN 0 11 551428 7
9. *Design and Analysis of Urban Storm Drainage; The Wallingford Procedure, Volume 4 - The Modified Rational Method,* National Water Council 1981
10. Highway Authorities and Utilities Committee. *The Specification for the Reinstatement of Openings in Highways,* 1992. HMSO
11. Roads and Street Works Act 1991 January 1993.
12. Transport and Road Research Laboratory. *TRRL Contractor Report 2, The Drainage Capacity of BS Road Gullies and a Procedure for Estimating Their Spacing,* 1985. TRRL

# 10 Road Pavements

Roads are essential for the economic and social well-being of our country; 89 per cent of all our freight is moved by road and we spend £6 billion a year on construction and maintenance. The need for reliable durable roads was realised in Britain during the Industrial Revolution and two men were instrumental in their provision, John Macadam and Thomas Telford. Telford's construction consisted of layers of hand pitched stone, decreasing in size towards the surface. Macadam's construction was thinner because he allowed for grading of stone in the respective layers, which increased interlock and thereby proved stronger at less cost. Both men realised the need for a foundation or capping layer and both realised the importance of drainage to maintain the strength of the road. With the advent of the motor car, dust proved to be a problem and tar was spread on the surface of the road to bind it together.

Today we generally use two types of road construction, flexible and rigid. Flexible pavements are constructed of granular material bound with some form of bitumen, tar or cement, whilst rigid pavements are constructed of concrete slabs. The choice between the two forms of construction is made at the design stage depending upon a number of factors. These include the design life, expected traffic flow and rate of growth, the strength of the subgrade, cost and ease of construction and the cost and ease of maintenance. Today 95 per cent of the new roads constructed in the UK are of flexible construction. This may be difficult to understand at first because the initial cost of a concrete road is less than that of a flexible road and it lasts longer, but at the present stage of development flexible roads are more reliable in terms of performance, easier to maintain and easier to construct.

Road pavement design is essentially about understanding the strength of the materials of which it is made and the strength of the ground on which it stands. Road pavement design can be carried out by the application of computer programs or by reference to the appropriate design standards which are in part based upon empirical test data. The computer program look at direct stress and strain calculations and are beyond the scope of this book. Road pavement design is at present based on the Department of Transport Design Standard HD 14/87[5] and the accompanying Advice Note HA 35/87[6] which is for motorways and major trunk roads. The Road Research Laboratory Report 1132[8] is used to consider the designs of lesser roads. In this chapter we will look at simple design standards only relying on frequent references to source text. We shall also consider road materials and look briefly at the construction and maintenance aspects.

## 10.1 THE PAVEMENT

There are a number of specific terms which apply to a road pavement and it may be prudent at this stage to run through them. A typical cross-section of a road pavement is shown in figure 10.1, giving the general terms used today. A capping layer is shown, but this is only necessary if the subgrade is weak.

Table 10.1

| Type of construction | Surfacing material | Roadbase material |
|---|---|---|
| Flexible construction | Bitumen binder | Bitumen binder |
| Composite construction | | |
| 1. Flexible composite | Bitumen binder | Cement binder |
| 2. Rigid composite | Bitumen binder | Concrete |
| Rigid construction | Concrete slab construction 1. Jointed unreinforced 2. Jointed reinforced 3. Continuously reinforced | |

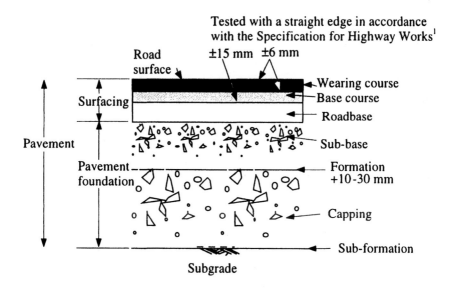

Figure 10.1

There are three basic types of road pavement, flexible construction, composite construction and rigid construction. The difference between them is the type of material from which the surfacing and roadbase are constructed. Flexible construction is where bitumen bound materials are used, composite is where bitumen and cement bound materials are used, and rigid is where concrete slabs are used. These differences are summarised in table 10.1.

The choice of the type of road construction is to be made at the design stage and depends upon the use to which the road is to be put and the strength of the ground on which it is built.

### 10.1.1 Flexible Construction

A typical general arrangement of a flexible construction is shown in figure 10.2. It shows surfacing comprising bitumen-bound wearing and base courses and a bitumen-bound roadbase which is the main structural layer. All the thicknesses shown in figure 10.2 will vary depending upon application and must be designed.

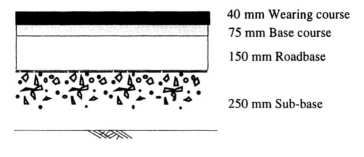

40 mm Wearing course

75 mm Base course

150 mm Roadbase

250 mm Sub-base

Figure 10.2

The advantages of this type of construction are that it is very quick to lay and is easy to maintain, but its initial construction costs are more expensive than the concrete equivalent. The Department of Transport have however, began to look at 'whole life costing' and if we take into account the ease of patching, resurfacing and partial reconstruction the economic arguments for concrete are eroded. Flexible construction is the most popular form of construction for motorways and trunk roads. Thinner overall construction thicknesses are used for surfacing in car parks, and lightly trafficked roads. Block pavements are sometimes used in place of surfacing in car park situations.

### 10.1.2 Composite Construction

There are two types of composite construction, flexible composite and rigid composite. Flexible composite is most widely used for road construction. It consists of a bituminous-bound surfacing and upper roadbase, where necessary, on a roadbase, or lower roadbase, of cement bound material. This is called 'Lean-mix' because it is a concrete of low compressive strength (7.5 to 15 N/mm²). The lean-mix usually takes the place of, and is the same thickness as, the granular sub-base layer, but it is recommended that some form of granular capping is used to cover the formation. This is to protect the formation from damage due to weather and the movement of construction vehicles whilst laying the lean-mix; the capping also aids drainage. The extra thickness of capping can be taken into consideration in the design and is allowed to count as part of the thickness of the sub-base. A general arrangement of a flexible composite is shown in figure 10.3.

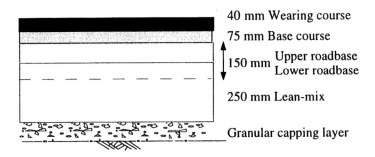

40 mm Wearing course

75 mm Base course

150 mm Upper roadbase
Lower roadbase

250 mm Lean-mix

Granular capping layer

Figure 10.3

Rigid composites are rarely used as a design choice but result from the overlay of surfacing on to a concrete road construction.

### 10.1.3 Rigid Construction

A typical general arrangement of a rigid construction is shown in figure 10.4. It shows combined surfacing and roadbase layers comprising a concrete slab which is the main structural layer. All the thicknesses shown in figure 10.4 and the size of the reinforcing mesh will vary depending upon application and must be designed. The concrete slab is constructed using an air-entrained concrete of 40 N/mm² strength.

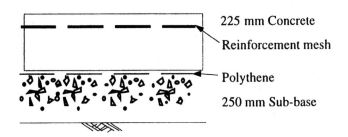

225 mm Concrete

Reinforcement mesh

Polythene

250 mm Sub-base

Figure 10.4

The air-entraining enables the concrete to resist frost attack. Concrete is very sensitive to thermal expansion and contraction and must therefore be constructed with contraction and expansion joints. Contraction and expansion joints are spaced at 4 and 40 m centres respectively for a slab thickness up to 225 mm and 5 and 60 m centres respectively for a slab over 225 mm thick. The polythene is provided to give a low friction surface between the concrete and the sub-base which encourages cracking at joint locations only and to make the construction watertight at the joints.

The advantages of this form of construction are that concrete is cheaper and more durable than bitumen-bound materials and in theory has a design life of 40 years without major reconstruction. In practice, however, it is usually the joints of the formation that fail before this time. The disadvantages are that concrete is slow to construct in comparison to bituminous-bound materials, it must be allowed to cure for up to 7 days before it can be used, and during that time must be protected from moisture loss and the weather. A concrete paving train is expensive to set up and work must be continuous to effect economies. Concrete is difficult to maintain, it cannot be patched like bituminous materials, and it is difficult to overlay because in time it separates from the concrete beneath.

## 10.2 ROAD BUILDING MATERIAL

The characteristics of the materials of which the pavement is constructed are one of the two major factors which determine the serviceable life, the other being the strength of the formation. For this reason a lot of effort is put into the testing and specifying of road materials. Specification of all materials used in the construction of roads can be found in the Department of Transport Manual of Contract Documents for Highway Works, Vol 1 and 2[7].

### 10.2.1 Capping Layer

This material is selected granular material usually, local stone ranging in size up to 125 mm and of a grading which is not frost susceptible, as shown in table 10.2. Capping layers are only used to reinforce a weak subgrade. The strength of the subgrade is measured by the California Bearing Ratio (CBR), see later.

### 10.2.2 Sub-base Material

The sub-base provides the upper layers of the pavement foundation and as such is structurally significant. It provides a regulating layer on which construction of the pavement can take place and protects the subgrade against damage from construction plant and the effects of weather. In the finished pavement the sub-base layer acts as a drainage layer to prevent the build-up of ground water. Any material within 450 mm of the road surface must be frost resistant. This means that the grading of the granular sub-base must be carefully controlled and conform to the requirements of the Department of Transports (DoT) Manual of Contract Documents for Highway Works, Vol 1 and 2[7].

Table 10.2

| BS sieve size | Capping 6F2 | Sub-base Type 1 | Sub-base Type 2 |
|---|---|---|---|
| 125 mm | 100 | - | - |
| 90 mm | 80 - 100 | - | - |
| 75 mm | 65 - 100 | 100 | 100 |
| 37.5 mm | 45 - 100 | 85 - 100 | 85 - 100 |
| 10 mm | 15 - 60 | 40 - 70 | 45 - 100 |
| 5 mm | 10 - 45 | 25 - 45 | 25 - 85 |
| 600 μm | 0 - 25 | 8 - 22 | 8 - 45 |
| 75 μm | - | 0 - 10 | 0 - 10 |
| 63 μm | 0 -12 | - | - |

*Extracted from the Manual of Contract Documents for Highway Work*
*Crown copyright is reproduced with the permission of the Controller of HMSO*

*Granular Sub-base*

The material most commonly used as a sub-base is granular and is divided into sub-base types 1 and 2. Grading is taken from the DoT Manual of Contract Documents for Highway Works, Vol 1 and 2[7], an extract of which is shown in table 10.2. It is usually a naturally occurring material which has had its grading characteristics altered to comply with DoT requirements; for example, this can be done by adding fine material, but sometimes the correct grading can be found naturally. As a 'rule of thumb' Type 1 usually consists of angular stones, whilst Type 2 consists of rounded stones. Type 1 material often consists of crushed rock graded to the DoT specification, whilst Type 2 is generally naturally occurring. When compacted, the granular sub-base provides a stiff dense layer which gives some protection against deformation and the weather.

*Lean-mix Concrete*

This is a very weak concrete of strength ranging from 7.5 to 15 N/mm². This material can be used in 'flexible composite construction' because a close network of cracks develops, when cured; this allows the material to behave just like interlocking stones. Lean-mix material is most widely used as a sub-base because it is strong, but as with all concrete materials, it is slow to construct because of a seven day curing period which must be allowed before the next layers can be laid. Also the surface of the lean-mix must be sealed with a tack coat of bitumen slurry whilst it cures to protect it from loss of moisture and help the adhesion of upper layers.

*Soil Cement*

This can be used if the subgrade is of a nature suitable to be mixed with cement. For example, if the subgrade consists of a granular material then it can sometimes be stabilised by mixing cement into the top surface. This creates a kind of in situ lean-mix. Lime stabilisation works on the same principle but in this case lime is mixed into a predominantly weak clay based material. This has to be done in the right quantities but has the affect of increasing the stiffness of the formation.

*Quarry Waste*

Quarry waste, sometimes called 'scalpings', can be used provided that the graded is acceptable. This material is a by-product of the grading operation for other materials. Obviously, this must be taken as found and may not conform to the Type 2 grading criteria.

*Hard-core*

Hard-core consists of broken bricks and building waste and does not conform to the DoT requirements and so cannot be used in road construction. The material may be used in private roads or car parks but not on estate roads that are to be adopted. The material can only be used in this situation if it is reasonably broken and does not contain steel, decomposable timber or other deleterious material. Brick hard-core is high in sulphate content which attacks concrete unless sulphate resisting cement is used, but even then sulphate levels must be monitored. Brick hard-core, therefore, is not recommended for use with rigid road construction due to the risk of sulphate attack against the concrete. There is also a recent European Community (EC) environmental directive which requires that brick rubble from demolished buildings must be cleaned and re-used for brick construction, instead of being broken up for hardcore; this makes the material scarce and expensive.

*Hoggin*

This is a mixture of clay and granular material and is not allowed for use by the Department of Transport in road construction. The granular material is 75 mm maximum sized stone of no specific grading mixed with a maximum of 20 per cent clay. This type of material is usually found naturally and can be used for private roads and car parks.

### 10.2.3 Geotextiles

Geotextile material is a mesh or fabric of polypropylene which is laid over the formation before the sub-base is laid. It is used in situations of a weak subgrade of less than 2 per cent CBR to improve the performance of the finished road. It acts in three ways. First, it prevents penetration of the sub-base into the formation under load; second, it confines lateral movement of the sub-base by tension in the geotextile, thus inhibiting rutting; and third, it allows excess water to pass into the sub-base to be drained away. Geotextiles have been used successfully in access roads, flexible and rigid pavement construction, embankment and railway construction. The Department of Transport, however, acknowledge only that the material has not been shown to improve the performance of a finished pavement.

### 10.2.4 Roadbase Material

This material is the main, load distributing layer in the construction and its thickness depends upon the traffic load expected, the type of material that it is composed of and the strength of the subgrade. The roadbase can be constructed of any of the following materials, a summary of which is shown in table 10.4.

*Dense Tarmacadam*

Dense tarmacadam to BS 4987[2] consists of 40 mm nominal size aggregate bound together with tar. The viscosity of the tar is denoted by the C number and for roads is usually in the range from C50 to C58. These values relate to the kinematic viscosity test as described in BS 76[4]. It is the temperature at which 50 ml of tar has a flow time of 50 seconds through a standard 10 mm orifice. The temperatures range from 38° to 58° C and directly relate to the C number. The higher the C number then the higher is the viscosity of the material at normal operating temperatures. C58 is suitable for a dense road base macadam for heavily trafficked roads, whereas C54 would be suitable for open textured macadam on a lightly trafficked road. Bitumen can be mixed with tar to improve the strength of the material but is not as robust as hot rolled asphalt or lean-mix concrete. Tarmacadam is cheaper than hot rolled asphalt and can be used in car parks or small estate roads.

*Dense Bitumen Macadam (DBM)*

DBM to BS 4987[2] is also 40 mm nominal size aggregate but bound together with bitumen. Again viscosity varies with traffic load (100/200 pen). These values relate to the Penetration Test. In this test a standard needle is allowed to penetrate a sample of bitumen, under a load of 100 gr at a fixed temperature of 25° C for a time period of 5 seconds. The depth of penetration determines the pen number. For example a penetration of 10 mm gives a pen number of 100. Thus the greater the penetration the softer the bitumen. Harder bitumen grades 35 to 100 pen are used in asphalt where stiffness is important, whilst softer grades 100 to 450 pen are used in macadams where lubrication and bonding of the aggregate are more important.

*Hot Rolled Asphalt*

Rolled asphalt to BS 594[1] is 40 mm aggregate bound together with asphalt (50 pen) and is stronger as a material than dense tarmacadam and DBM. It is often used to provide a water-resistant layer within the pavement design if it is not provided by an impermeable surface. This is an expensive option but has the advantage of speed of construction.

All of the above materials are strong due to their density, flexible due to the nature of the binder and frost resistant. Different viscosity denoted by different pen or C numbers means that the materials must be laid at different temperatures. It is, therefore, important that successive layers should contain binders of similar viscosity so that they may be laid at compatible temperatures.

### 10.2.5  Base Course Material

This material is required to distribute traffic loads over the roadbase and provide a well shaped regular surface on which to lay the relatively thin wearing course. The material used depends upon the traffic loading as shown in table 10.4.

*Hot Rolled Asphalt*

Rolled Asphalt base course 50 pen to BS 594[1] is the same material as used in the asphalt roadbase except that it has a higher binder content and smaller stone sizes. Aggregate size depends on layer thickness, as shown in table 10.3.

Table 10.3

| Stone size | Thickness |
|---|---|
| 20 mm - 35 mm | 50 mm thick |
| 28 mm - 50 mm | 60 mm thick |
| 40 mm - 60 mm | 80 mm thick |

*Dense Coated Macadams*

Dense coated macadams (dense bitumen macadam 100 pen and dense tarmacadam C50 to C58) to BS 4987[2] have stone sizes as indicated by table 10.3 but there is less binder. A 'macadam' material relies for its strength on the interlocking of the aggregate content, so grading is very important.

### 10.2.6  Wearing Course Material

This material is required to provide a regular shaped riding surface and to withstand direct tractive forces and loads from traffic. It seals the road structure against weather and provides a durable, skid resistant surface. The material used depends upon traffic loading as shown in table 10.4.

*Hot Rolled Asphalt*

Rolled asphalt to BS 594[1] has a coarse aggregate content (crushed rock or slag) of 30 % consisting of 10 mm single sized stone bound together with asphalt (50 pen) and laid to a minimum thickness of 40 mm. Chippings of 20 or 14 mm, coated with mastic asphalt are rolled into the surface when still hot to provide a very durable skid resistant surface. Rolled asphalt is currently the only material specified by the DoT for the surfacing of trunk roads and motorways of flexible construction.

Table 10.4

| Loading | 10-100 cv/d | 100-3000 cv/d | 3000-1000 cv/d |
|---|---|---|---|
| Wearing course | Hot rolled asphalt (40 mm thick) Dense tarmacadam Dense bitumen macadam Block pavements* | Hot rolled asphalt (40 mm thick) | Hot rolled asphalt (40 mm thick) |
| Base course | Hot rolled asphalt 50 pen Dense bitumen macadam Dense tarmacadam Block pavements* | Hot rolled asphalt Dense bitumen macadam Dense tarmacadam (60 mm thick) | Dense bitumen macadam (125 mm thick) |
| Road-base | Dense bitumen macadam Dense tarmacadam Hot rolled asphalt | Dense Tarmacadam Dense bitumen macadam Hot rolled asphalt | Hot rolled asphalt |
| Sub-base | Granular sub-base Lean-mix concrete Soil cement | Sub-base Type 1 (Type 2 < 400 cv/d) Lean-mix | Sub-base Type 1 Lean-mix |
| Capping | Local material | Local material | Local material |

*cv/d is commercial vehicles per day*
*\* Block Pavements are only used for car parks in place of the Wearing Course and Base Course thicknesses.*
*Crown copyright is reproduced with the permission of the Controller of HMSO*

*Dense Bitumen Macadam*

Dense bitumen macadam surfacing to BS 4987[2] has a stone content of 30-50 % of 10 mm size bound together with bitumen (100 pen). This is commonly used for estate roads.

*Dense Tarmacadam*

Dense tarmacadam surfacing to BS 5273[3] has a stone content of 30-50 % of 10 mm size bound together with tar. This is not for heavily trafficked areas but is resistant against oil spills and is widely used in car or lorry parks and also for estate roads.

*Open Textured Tarmacadam*

Open textured tarmacadam surfacing (bitumen macadam or tarmacadam) to BS 4987[2] has a nominal stone size of 10 mm and can be used in a thickness of 20-25 mm. However, it does not last long and must be surface dressed within 2 years. Suitable for footpaths. Medium textured macadams are suitable for footpaths and light traffic.

*Block Pavement*

Block pavements consist of concrete blocks about 100 mm wide 200 mm long and 65 to 80 mm thick placed on a 20 mm bed of sand. In the construction and design the combined thickness can take the place of base course and wearing course.

## 10.3 PAVEMENT DESIGN

Pavement design involves the assessment of the strength of the subgrade, an assessment of the possible traffic loading and matching these to the best type of materials and thicknesses.

### 10.3.1 Subgrade

The best method at the moment of testing the strength of a subgrade is the CBR test (California Bearing Ratio). There are three ways of doing this.

- in the laboratory
- in situ plate bearing test
- by soil classification

In the laboratory a premeasured mass of material is recompacted into a cylinder mould. A metal plunger 45 mm in diameter is pushed into the soil sample at a rate of 1.08 mm per minute. The moisture content must be equal to that which is expected in situ and the force required to push the plunger in is plotted against the penetration. This test is compared as a percentage of the standard curve (carried out on crushed rock) to give the CBR value as shown in figure 10.5.

The in situ plate bearing test can be carried out by hydraulic rams pushing a plate 300 mm square into the soil under the weight of a lorry. The plate is at the bottom of a freshly dug 1m deep hole in an attempt to reproduce the condition that can be expected at the formation level under a road.

A simplified soil classification chart is shown in table 10.5. This relates Plastic Index to CBR values and gives a good overview of values. The gravel samples are further classified by grading. The chart gives values allowing for the depth to water table, the thickness of pavement construction including capping and allowing for average construction conditions. For example, if construction conditions are poor due to excessive ground or surface water then CBR values must be reduced by approximately 50 per cent for clay and silt. If they are good, they can be increased by about 10 per cent. Further information can be gained from TRRL Report 1132[8].

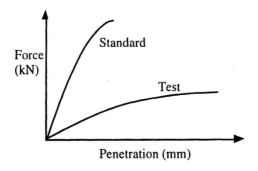

Figure 10.5

The strength of the subgrade depends on adequate drainage. It is very important to drain the subgrade so that it is not softened by excess ground or surface water. Drainage must be provided to keep the water table at least one metre below the formation level. If this cannot be done then the CBR must be adjusted as shown in table 10.5.

Other precautions to be taken against degradation of the formation are that it should not be exposed to weather during construction and must be covered with sub-base as soon as possible.

Impermeable road constructions must also be used in the final state to help protect the subgrade. An asphalt, dense bitumen macadam or concrete material must be used in one of the pavement layers to provide an impermeable layer to protect the subgrade from surface water that percolates down through the pavement construction. Tarmacadam may not be considered as 'waterproof'.

## 10.3.2 Frost Susceptibility

It is important that the subgrade be assessed for frost susceptibility. The effect of frost on clay is a large expansion in volume known as frost heave and when the frost has thawed the subgrade is left with much-reduced shear strength and CBR value. Chalk is dramatically affected by frost, both in terms of expansion and degradation. In the UK frost rarely penetrates the ground deeper than 450 mm and so materials used in the upper 450 mm of construction should not be frost susceptible. Road pavements constructed on chalk must always be designed to the minimum thickness of 450 mm despite the fact that the CBR values may well indicate that a thinner construction is possible.

Table 10.5 Equilibrium CBR values for
average construction conditions related to soil classification

| Soil type | Plastic index | Water table 300 mm beneath formation | | Water table 1000 mm beneath formation | |
|---|---|---|---|---|---|
| Construction thickness including capping* (mm) | | 300 | 1200 | 300 | 1200 |
| Heavy CLAY | 70 | 2 | 2 | 2 | 2 |
| | 60 | 2 | 2 | 2 | 2 |
| | 50 | 2 | 2.5 | 2 | 2.5 |
| | 40 | 2.5 | 3 | 3 | 3 |
| Silty CLAY | 30 | 3 | 4 | 4 | 4 |
| Sandy CLAY | 20 | 4 | 5 | 5 | 6 |
| | 10 | 3 | 6 | 4.5 | 7 |
| SILT** | - | 1 | 1 | 2 | 2 |
| Poorly graded SAND | - | 20 | 20 | 20 | 20 |
| Well graded SAND | - | 40 | 40 | 40 | 40 |
| Well graded sandy GRAVEL | - | 60 | 60 | 60 | 60 |

*Extract from Table C1 in TRRL. Report 1132, courtesy of the Road Research Laboratory*
*\* CBR values may be interpolated between 300 mm and 1200 mm thickness*
*\*\* Estimated assuming some probability of material saturation*

### 10.3.3 Loading

Loading is expressed in terms of the numbers of commercial vehicles expected to use the road per day (cv/d) at the time of opening (cars do not count). The average number of commercial vehicles travelling one way in any 24 hour period is called the annual average daily flow (AADF). This is usually estimated from a traffic flow analysis. A growth rate of 2 per cent is applied to this estimate and is built into design charts provided by the DoT and TRRL. Table 10.6 is taken from Road Note 29[9] a document which has now been superseded by HD 14/87[5], but it gives a good estimate of opening traffic flow in the absence of any other data. Design life is usually taken as 40 years.

### 10.3.4 Design

The required thicknesses of pavement foundation, sub-base and capping are given in figure 10.6. These requirements should be applied to all pavement construction from motorways to car parks, 10 cv/d to 10,000 cv/d. The thicknesses of roadbase and surfacing for a flexible construction and a flexible composite construction are given in tables 10.6, 10.8 and 10.9. These figures may not be interpolated. In table 10.6 construction thicknesses are given in terms of loading, subgrade strength and design life. A design life of 20 and 40 years is given because such construction may be carried out for private Clients who do not wish to pay for the specification of 40 years as laid down by the DoT. It also demonstrates how small an increase in construction thickness is necessary to double the design life. Tables 10.7 relate loading to the type of road; the pavement foundation thickness is shown in figure 10.6.

*Example*

Design a flexible road construction for the following conditions CBR 3 per cent and loading is estimated at 250 commercial vehicles per day.
From figure 10.6, sub-base thickness is 150 mm with 350 mm of capping.

From table 10.8, for a loading of 400 cv/d a roadbase thickness of 160 mm is required with a base course of 60 mm and surfacing of 40 mm thickness.

Form table 10.4, we can select our materials: hot rolled asphalt surfacing, with dense bitumen macadam base course and roadbase, Type 2 granular sub-base and a capping layer of local material.

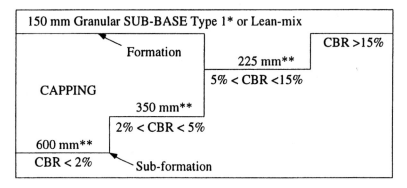

Figure 10.6 Flexible and Flexible Composite pavement foundation thickness
*\* Type 2 Sub-base may be used for loading less than 400 cv/d.*
*\*\* Dimensions shown relate to the thickness of the layers referred to.*
*Note. If the subgrade is frost susceptible then the total thickness of the pavement and pavement foundation should not be less than 450 mm.*
*Crown copyright, reproduced with the permission of the Controller of HMSO*

Table 10.6 Flexible Road Construction Thicknesses (10 to 100 cv/d)

| Loading (cv/d) | 100 | 75 | 50 | 10 |
|---|---|---|---|---|
| Design life (years) | 20/40 | 20/40 | 20/40 | 20/40 |
| Cumulative number of commercial vehicles at 2% growth ($\times 10^6$) | 1.75/3.60 | 0.80/2.70 | 0.55/1.80 | 0.15/0.38 |
| Cumulative standard axles ($\times 10^6$)* | 0.47/1.62 | 0.36/1.22 | 0.25/0.81 | 0.07/0.17 |
| **CBR 5%** | | | | |
| sub-base | 225 mm of granular type 2 or 150 mm of type 2 on 600 mm of capping | | | |
| Roadbase thickness** | 30/90 | 30/90 | 30/60 | 10/20 |
| **CBR 2%** | | | | |
| sub-base | 150 mm of type 2 on 600 mm of capping | | | |
| Roadbase thickness | 70/140 | 70/140 | 60/90 | 20/60** |
| Surfacing | 40 mm wearing course and 60 mm base course | | | |

* *using a conversion factor of 0.45 × cumulative number of commercial vehicles.*
** *these thicknesses may be incorporated into the base course layer.*
*These figures may not be interpolated.*
*Crown copyright, reproduced with the permission of the Controller of HMSO*

Table 10.7 Commercial Traffic Flows
Recommended for design in the absence of any other data.

| Type of road | Estimated traffic flow of commercial vehicles per day in each direction at the time of construction |
|---|---|
| Cul-de-sacs and minor residential roads | 10 |
| Through-roads and roads carrying regular bus routes involving up to 25 public service vehicles (PSV) per day in each direction | 75 |
| Major through roads carrying regular bus routes involving 25-50 PSVs per day in each direction | 175 |
| Main shopping centre of a large development carrying goods deliveries and main through-roads carrying more than 50 PSVs per day in each direction | 350 |

*Crown copyright, reproduced with the permission of the Controller of HMSO*
*Note, Table 10.6 is taken from Road Note 29[9] which has been superseded by TRRL.*
*Report 1132[3] and may not necessarily represent current best practice for the*
*construction and maintenance of trunk roads.*

Table 10.8 Thickness of Roadbase for Flexible Road Construction

| Loading (cv/d) | 100 | 200 | 400 | 800 | 1000 | 2000 | 3000 | 6000 | 10,000 |
|---|---|---|---|---|---|---|---|---|---|
| Roadbase thickness (mm) | 90 | 120 | 160 | 210 | 220 | 270 | 225 | 265 | 275 |
| Surfacing | 40 mm wearing course and 60 mm base course | | | | | | 40 mm wearing course and 125 mm base course | | |

*cv/d is commercial vehicles per day*
*Information extracted from chart 3 of HD 14/87[1]*
*More accurate values can be obtained by direct reference to HD 14/87[1]*
*Crown copyright, reproduced with the permission of the Controller of HMSO*

Table 10.9 Thickness of Roadbase for Flexible Composite Road Construction

| Loading (cv/d) | 100 | 200 | 400 | 800 | 1000 |
|---|---|---|---|---|---|
| Wearing course thickness (mm) | 40 | 40 | 40 | 40 | 40 |
| Base course thickness (mm) | 60 | 60 | 80 | 100 | 110 |
| Roadbase thickness (mm) | 150 | 150 | 180 | 230 | 250 |

*cv/d is commercial vehicles per day*
*Information extracted from chart 4 of HD 14/87[1]*
*More accurate values can be obtained by direct reference to HD 14/87[1]*
*Crown copyright, reproduced with the permission of the Controller of HMSO*

## 10.4  CONSTRUCTION

The pavement foundations are constructed using the same plant and compaction techniques as used in chapter 5 except that the material will generally come in by road lorries because suitably graded materials do not occur naturally on site. The granular materials for capping and sub-base are graded and dozed with the grader. The material is compacted in accordance with table 5.5 of chapter 5. Much more care is needed when laying bitumen bound materials and different plant is needed.

In general, bitumen-bound materials are laid using a paving machine as shown in figure 10.7, although small areas may be laid by hand. The function of the paving machine is to lay an even thickness of material in strips usually in widths of 2.5 to 3 m. It lays the material at the right temperature and tamps it ready for compaction by the roller. The range of width setting for the machine is between 1.5 and 6 m. The principle of operation is that insulated lorries deliver the material into the hopper and it is taken by conveyor to the spreading augers. The augers spread the material across the width required and it is then tamped to an even surface by the screedboard and vibrotamper. The thickness of the material laid can be controlled manually or automatically by the use of sensors attached to an averaging bar mounted along the side of the machine.

Bitumen and tarmac-based materials will not spread or compact correctly unless they are used at the right temperature, which is between 95° to 190° C depending upon the grade of binder. The material loses about 5° C during the laying process and therefore must arrive on site at the correct temperature. The vibrotamper is heated to aid laying to a temperature about 5° C above the temperature of the material being laid. Temperature checks should be made throughout the laying process. It is of absolute importance that the flow of material to the augers is not interrupted because this will cause an uneven surface and poor compaction. The Contractor must therefore ensure that there is available an uninterrupted supply of material at the correct temperature and never start laying with less than two supply lorries on site. For hot rolled asphalt surfacing, bitumen-coated chippings are sprinkled onto the uncompacted surface by a chipping spreader as shown in figure 10.8. This machine travels along directly after the paving machine.

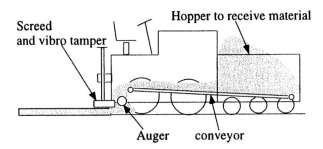

Figure 10.7 Paving machine

Correct rolling is very important to the quality of the finished surface. Rolling is carried out with an 8 to 10 tonne dead-weight roller or a tandem vibratory roller. The dead-weight roller has the advantage of allowing rolling to take place at higher temperatures and gently pushes the material into its compacted position but it does not compact the edges very well. The tandem vibratory roller has the advantage of being able to compact the edges well but can be too violent with uncompacted material. The best results will be achieved if both are used, dead weight first and then the vibratory tandem, then the advantages of both machines can be used. It is important that rolling takes place at the right temperature; too hot and the material will be displaced too much producing an uneven finish, too cold and the material will not be adequately compacted. Rolling techniques are also important. It has been shown that compaction is the most important factor in ensuring satisfactory design life. The dead weight roller should always work with its larger wheels toward the paving machine to reduce pushing of the uncompacted material and the roller should travel at a slow steady rate. Points at which the direction of travel is reversed should be staggered and changing of lateral position by steering must only be done on compacted material. The lower edge should be rolled first followed by the other edge and then the middle, working over the area from the lower edge in half-wheel widths.

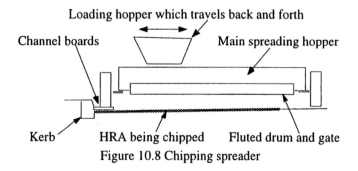

Figure 10.8 Chipping spreader

### 10.4.1 Road Maintenance

A flexible road has a design life of 40 years. A flexible road construction is expected to be resurfaced at 10 years, overlaid or partially reconstructed at 20 years, resurfaced again at 30 years and completely reconstructed at 40 years. Because of the difficulties of resurfacing rigid road construction the surface is designed to last 40 years. There are a number of methods used to maintain roads, such as patching, resurfacing, overlaying, partial reconstruction and total reconstruction. Patching is not allowed on major trunk roads and motorways so the first remedial action is that of resurfacing or overlaying. The DoT recommends that overlay is carried out before the rutting becomes too severe and stipulates a critical condition beyond which major maintenance work may be required. The critical condition is when 10 mm deep ruts or longitudinal cracks are visible along 15 per cent of the inspected length of road. This work can be carried out by simply adding another layer of material to the road but tends to cause problems with kerb and manhole cover levels. It is more common to mill off the top surface with a large milling machine which is capable of removing up to 75 mm of road in one pass; new material is then laid to give the road a new running surface.

Rigid roads are difficult to maintain because resurfacing material does not bond well to the original surface. A common mode of failure is that slabs tilt with subgrade failure and must either be completely replaced to allow subgrade improvements or must be pressure grouted. Both operations are slow, expensive and do not give a consistent standard of repair. A new method open to Engineers is to shatter the concrete slab and use it as a sub-base or roadbase for a flexible composite construction. Base course and wearing course can be overlaid in the conventional manner, but this cannot be done on a piecemeal basis.

### 10.5 REFERENCES

1.  British Standards Institution. *BS 594 Pts 1 and 2:1992:Specification for the Transportation, laying and Rolling of Rolled Asphalt.*
2.  British Standards Institution. *BS 4987:Pts 1 and 2:1993, Coated Macadams*

*for Roads and Other Paved Areas.*

3. British Standards Institution. *BS 5273:1990: Specification. Dense Tar Surfacing for Roads and Other Paved Areas.*

4. British Standards Institution. *BS 76:1974: Specification for Tars for Road Purposes*

5. Department of Transport. *Design Standard HD 14/87, Structural Design of New Road Pavements, 1987,* (Including Amendment No. 1 dated 1988) HMSO.

6. Department of Transport. *Advise Note HA 35/87, Structural Design of New Road Pavements, 1987,* HMSO.

7. Department of Transport. *Manual of Contract Documents for Highway Works. Vol 1 Specification for Highway Works and Vol 2 Notes for Guidance on the Specification of Highway Works,* 1994. HMSO ISBN 0-11-55103-0 and ISBN 0-11-55103-1

8. Powell, Potter, Mayhew and Nunn, *The Road Research Laboratory Report 1132, Structural Design of Bituminous Roads,* 1984, The Road Research Laboratory, HMSO.

9. The Road Research Laboratory. *Road Note 29, A Guide To The Structural Design Of Pavements For New Roads,* 1970, HMSO.

# Index